Yasuhiro Koike

Fundamentals of Plastic Optical Fibers

Related Titles

Schnabel, W.

Polymers and Electromagnetic Radiation

Fundamentals and Practical Applications

2014
ISBN: 978-3-527-33607-4
Also available in digital formats

Nicolais, L., Carotenuto, G. (eds.)

Nanocomposites

In Situ Synthesis of Polymer-Embedded Nanostructures

2014
ISBN: 978-0-470-10952-6

Reed, G.T. (ed.)

Silicon Photonics - The State of the Art

2008
ISBN: 978-0-470-99453-5

Hadziioannou, G., Malliaras, G.G. (eds.)

Semiconducting Polymers

Chemistry, Physics and Engineering

Second Edition
2007
ISBN: 978-3-527-31271-9

Schnabel, W.

Polymers and Light

Fundamentals and Technical Applications

2007
ISBN: 978-3-527-31866-7
Also available in digital formats

Steinbüchel, A. (ed.)

Biopolymers

10 Volumes + Index

2003
ISBN: 978-3-527-30290-1

Iizuka, K.

Elements of Photonics, 2-Volume Set

2002
ISBN: 978-0-471-41115-4

Saleh, B.E., Teich, M.C.

Fundamentals of Photonics

1991
ISBN: 978-0-471-83965-1

Yasuhiro Koike

Fundamentals of Plastic Optical Fibers

Verlag GmbH & Co. KGaA

The Author

Yasuhiro Koike
Professor, Keio University
Director, Keio Photonics Research Institute
Yokohama, Japan

Cover

The picture represents how light propagates through graded index POFs. The incident light draws a sine curve because GI POFs have parabolic refractive index profiles in the core regions.

Library of Congress Card No.: applied for

British Library Cataloguing-in-Publication Data
A catalogue record for this book is available from the British Library.

Bibliographic information published by the Deutsche Nationalbibliothek
The Deutsche Nationalbibliothek lists this publication in the Deutsche Nationalbibliografie; detailed bibliographic data are available on the Internet at http://dnb.d-nb.de.

Print ISBN: 978-3-527-41006-4
ePDF ISBN: 978-3-527-64653-1
ePub ISBN: 978-3-527-64652-4
Mobi ISBN: 978-3-527-64651-7
oBook ISBN: 978-3-527-64650-0

Cover-Design Adam-Design, Weinheim, Germany
Typesetting Laserwords Private Limited, Chennai, India
Printing and Binding Markono Print Media Pte Ltd., Singapore

Printed on acid-free paper

Contents

Preface

People in the industry formerly held the vague notion that polymers were unsuitable for application in the high-performance photonics field. Twenty odd years later, however, we have seen the birth of photonic polymers in applications such as the world's fastest plastic optical fiber (POF) and high-resolution displays. Research papers written in the first half of the twentieth century by Einstein and Debye, which delve into the essence of light scattering, have become my personal bibles. From them, I learned that the more we strive to achieve a breakthrough, the more important it is that we return to the fundamentals.

What originally drew me into the academic field of photonic polymers, a field which combines photonics and physical sciences, was my meeting with the late Professor Yasuji Otsuka, a person who I have the greatest respect for. I was a fourth year undergraduate student when I joined Professor Otsuka's laboratory in 1976. At that time, the laboratory had just begun its research into optically converging plastic rod lenses. By creating a graded index (GI) in the radial direction in a rod-shaped polymer, the light passing through it would travel in a meandering path, forming an image even when both ends of the rod were flat. I found this to be a most curious phenomenon and became fascinated by polymers. What interested me at first was why a GI causes light to bend gradually according to the refractive index profile. Professor Otsuka was an expert in polymer chemistry, focusing on emulsion polymerization, but he worked out and explained Maxwell's ray equation to me. I was very impressed and also surprised at the fact that he had gone beyond his particular field of polymer chemistry and was attempting to use optics and mathematics to explain the behavior of light in a polymer. I greatly admired Professor Otsuka.

Years later, I would find myself researching the fundamental reasons for light scattering loss – the prime reason why fibers could not be made transparent. While I remained a member of the polymer chemistry laboratory, on my own I entered the worlds of physics and optics as I studied subjects such as light scattering theory and polarization. I would continue to learn in the academic field of photonic polymers for the next 20 years, but my fundamental research method was the one that I learned at that time from Professor Otsuka.

In 1982, I received Ph.D. from Keio University and I was at a crossroads in my research into the question of whether it was truly possible to create a GI-type POF that could transmit optical signals at speeds exceeding one gigabit. The prototype GI-POF suffered from low transparency, and had a transmission loss of more than 1000 dB/km, preventing the light from traveling farther than a few meters. In order to achieve transmission speeds exceeding one gigabit, it was necessary to create a GI by adding another material inside the fiber and creating a density distribution in the radial direction. However, at the same time, determining how to eliminate impurities in order to make the POF transparent was a major issue. Therefore, attempting to achieve high-speed optical communication by adding another material (an impurity?) that would create a GI was a large challenge.

At that time, in order to create a gradient index, two monomers M1 and M2 with different reactivities were copolymerized. A polymer containing large quantities of the highly polymerizable M1 monomer with its low refractive index was gradually deposited around the periphery to create the gradient index. The created GI preform appeared transparent, but it was possible to see the light beam traveling in a pretty meandering path which was visible as a result of scattering of the light. We needed to reduce this light scattering loss by a factor of at least several hundreds. I was completely unaware of the large and theoretically impenetrable barrier which lays in the way of continuously improving the copolymerzation method which uses the different reactivities of the monomers to create the gradient index, and I continued to work toward creating a transparent GI-POF.

The problem of light scattering consumed me, and I was not able to assign this as a research theme for student graduation or master's theses because I did not know whether the research would produce any results. I continued to struggle with the problem on my own. As I was reading a broad range of documents on the subject, I came across Einstein's fluctuation theory of light scattering from the early twentieth century. This was based on micro-Brownian motion in solution, and it proposes that light scattering loss is proportional to isothermal compressibility. When I actually entered the isothermal compressibility of the POF material PMMA (poly(methyl methacrylate)), I obtained a value of 10 dB/km – much lower than the aforementioned value of 1000 dB/km. This meant that transmission over a distance exceeding 1 km was possible. This was truly a revelation to me at that time. However, in that case, where did the difference of 990 dB/km come from? In order to identify the kind of heterogeneous structure which was producing this excessive scattering, I learned everything I could about light scattering using Debye's correlation function from the 1950s. I also tried again to work the theory out on my own. This was an extremely useful theory for quantitative analysis of the relationship between micro nonuniform polymer structures and light scattering. What was wonderful about Debye's scattering theory was that it was possible to define the correlation function without hypothesizing the shape or size of the heterogeneous structures in the polymer chain. This makes it possible to experimentally find the correlation distance that includes information about the heterogeneous structure shape and size from the angular dependence of the light scattering. This became a powerful tool for me in working to identify the cause of excess polymer scattering.

At the same time, I began to see how the method of forming a gradient index according to differences in reactivity would form an extreme polymer compositional distribution in the generated copolymer. It became clear that, as we tried to make the gradient index larger by increasing the difference in reactivity, the polymer composition would contain increasing amounts of components that were similar to homopolymers of M1 and M2. This essential large heterogeneous structure reached sizes in excess of several hundred angstroms, and when applied to Debye's light-scattering theory, I discovered it to be the cause of an enormously large scattering loss exceeding several hundred decibels per kilometer. This realization was the culmination of many long years of research and allowed me to recognize that the fiber would theoretically not become transparent utilizing processing methods that rely on monomer reactivity.

I began experiments based on the completely new idea of forming a gradient index using the sizes of the molecules instead of the monomer reactivity. I still remember the day very clearly. It was April 1, 1990 (April Fool's Day). On that day, I had produced a superbly transparent GI-POF preform. This was the moment I emerged from my 10-year-long search for a solution to scattering loss. Our laboratory was galvanized from this point on. Our course was clearly set, and all student research themes were channeled in this direction. Test data on GI-POF with increasingly lower losses and higher speeds were produced one after another. I obtained a patent, wrote numerous papers, and began joint research with industry all at once. News that an optical signal with speed exceeding a gigabit passed through 100 m of GI-POF for the first time was reported on August 31, 1994, on the front page of the Nikkei Shimbun (newspaper).

The production method up until around 2005 was the preform method (a manufacturing method where a GI preform is created, which is then made into GI-POF through heat-drawing) with a focus on the interfacial-gel polymerization method. However, from around 2005, we began to develop the continuous extrusion method in earnest, and by 2008 succeeded in 40-Gb transmission. This was the world's fastest transmission speed, surpassing the GI-type silica optical fiber. We achieved these results through joint research with Asahi Glass. It was based on the essential principle of the materials, which indicates that a perfluorinated polymer used as the POF core material has a lower material dispersion compared to silica. (Material dispersion determines the transmission bandwidth.)

My research into light scattering, which I have described here, has been supported by the research papers that were written by Einstein and Debye during the first half of the twentieth century, and which delve into the essence of light scattering. Through these papers, I realized that the latest research papers would not always be useful in pursuing leading-edge research. I learned that the more we strive to achieve a breakthrough, the more important it is that we return to the fundamentals.

Yasuhiro Koike

Acknowledgments

This book was written by Yasuhiro Koike with inputs from Kenji Makino (Chapters 1, 3, 6, and 7), Azusa Inoue (Chapters 2 and 3), and Kotaro Koike (Chapters 2, 4, and 5), Project Assistant Professors of Keio Photonics Research Institute, Keio University, Japan. Their works were supported by the Japan Society for the Promotion of Science (JSPS) through its "Funding Program for World-Leading Innovative R&D on Science and Technology" (FIRST Program).

1
Introduction: Faster, Further, More Information

The realization of these three features has motivated the development of communication systems since the dawn of history. Optical communication systems in the broad sense date back to ancient times. One of the earliest optical communication systems was fire and smoke. Although atmospheric conditions, such as rain, snow, fog, and dust, strongly affect the transmission reliability, this type of optical communication was used for a long time worldwide. In addition to the sensitivity to the environmental conditions, the signal receiver was the human eye; thus, the transmission system had poor reliability. More stable and dependable communication systems were developed; for instance, a courier or pigeon carried messages and letters.

The era of electrical communication started in 1837 with the invention of the telegraph by Samuel F. B. Morse. The telegraph system used the Morse code, which represents letters and numbers by a coded combination of dots and dashes. The encoded symbols were conveyed by sending short and long pulses of electricity over a copper wire at a rate of tens of pulses per second. The telegraph and Morse code dramatically improved the speed, quality, and information capacity of transmission, although well-trained and skilled operators were required.

Another giant leap in the history of communication systems was brought about by Alexander Graham Bell in 1876. Bell developed a fundamentally different and user-friendly device that could transmit the entire voice as is, in an analog signal. The device is the telephone, which rapidly increased the speed and quality of communication. Subsequently, the invention of the facsimile machine enabled the transmission of figures and drawings. The development of electrical communication systems shifted to use progressively higher frequencies, which offered increases in bandwidth or information capacity. Optical communication was gradually becoming attractive because optical frequencies are several orders of magnitude higher than those used by electrical communication systems. Therefore, the optical carrier frequencies yield a far greater potential transmission bandwidth than electrical systems with metallic cables.

No significant advance in optical communication appeared until the invention of the laser in the early 1960s because no practical transmitter existed, and all communication systems must include the fundamental elements of a transmitter, the transmission medium, and a receiver. The invention of the laser aroused

Fundamentals of Plastic Optical Fibers, First Edition. Yasuhiro Koike.

curiosity about the possibility of using the optical wavelength region of the electromagnetic spectrum for transmission systems. The optical frequencies generated by such a coherent optical source are on the order of 10^{14} Hz, and the laser theoretically has an information capacity exceeding that of microwave systems by a factor of 10^5. Experimental investigations using atmospheric optical channels were conducted in the early 1960s with the potential of such broadband transmission capacities in mind. However, the transmission quality was unstable, again depending on the atmospheric conditions. At the same time, it was recognized that an optical fiber can provide a more reliable transmission channel because it is immune to environmental conditions [1]. The initial optical fibers appeared impractical because of their extremely large optical losses of more than 1000 dB/km. The situation changed in 1966, when Kao and Hockham speculated that the high losses were a result of impurities in the fiber material, and that the losses could potentially be reduced significantly in order to make optical fibers a functional transmission medium [2]. The technical breakthrough for optical communication occurred in 1970, only 4 years after this prediction; an optical fiber was demonstrated with a purified silica glass having an optical power loss low enough for a practical transmission link [3]. Charles K. C. Kao won the Nobel Prize in physics in 2009 for his pioneering insight and his enthusiastic further development of low-loss optical fibers. Ultimately, these efforts resulted in practical optical communication systems widely used across the earth. Optical fiber systems commonly provide great advantages such as longer distance, greater information capacity, immunity to electromagnetic interference, information security, smaller size, and lighter weight. Applications for these sophisticated networks include web browsing, e-mail exchange, telemedical care, remote education, and grid and cloud computing.

The telecommunication industries seriously considered standardizing on multimode fibers (MMFs) before deciding to adopt single-mode fibers (SMFs; see the next section). MMFs are very attractive because the mechanical tolerances of link components, such as connectors, splices, and coupling optics, are remarkably relaxed compared to those of SMFs. However, MMFs were considered unsuitable for telecommunication systems in the final analysis because of two critical technical problems: modal noise, and unstable bandwidth performance. Modal noise is attributed to fluctuation in time of the speckle pattern resulting from interference between the propagating modes, and bandwidth instability arises from the central dip in the refractive index. In contrast, plastic optical fibers (POFs) have an enormous number of propagating modes, and hence speckle patterns, which practically cancel the modal noise effect by averaging the fluctuations. POFs have essentially no central dip because of fabrication processes. Thus, serious modal noise and unstable bandwidth have not been observed, even though POFs are MMFs. Therefore, because the large mechanical tolerance provides easy and cost-effective installation, MMFs, particularly POFs, have become attractive again and have gradually been installed in short-reach networks such as local area networks (LANs) in homes, offices, hospitals, and vehicles, and even very short-reach networks such as interconnects inside computers which contain many connections.

This chapter briefly describes the fundamentals of optical fibers and provides an introductory review of the development of POFs.

1.1 Principle of Optical Fiber

The principle of light propagation through optical fibers is simply explained as follows [4, 5]. An optical fiber generally consists of two coaxial layers in cylindrical form: a core in the central part of the fiber and a cladding in the peripheral part that completely surrounds the core. Although the cladding is not required for light propagation in principle, it plays important roles in practical use, such as protecting the core surface from imperfections and refractive index changes caused by physical contact or contaminant absorption, and enhancement of the mechanical strength. The core has a slightly higher refractive index than the cladding. Therefore, when the incident angle of the light input to the core is greater than the critical angle determined by Snell's law, the input light is confined to the core region and propagates a long distance through the fiber because the light is repeatedly reflected back into the core region by total internal reflection at the core–cladding interface. The propagation of light along the fiber can be described in terms of electromagnetic waves called *modes*, which are patterns of electromagnetic field distributions. The fiber can guide a certain discrete number of modes that must satisfy the electric and magnetic field boundary conditions at the core–cladding interface according to its material and structure and the light wavelength.

Optical fibers can be commonly classified into two types: SMFs and MMFs [6]. As the names suggest, SMF allows only one propagating mode, whereas MMF can guide a large number of modes. Both SMFs and MMFs are again divided into two classes: step-index (SI) and graded-index (GI) fibers. The SI fiber has a constant refractive index in the entire core. The refractive index changes abruptly stepwise at the core–cladding boundary. The GI fiber has a nearly parabolic refractive index distribution. The refractive index decreases gradually as a function of the radial distance from the core center. Figure 1.1 conceptually illustrates the refractive index profiles and ray trajectories in SMF and in SI and GI MMFs, and shows the measured input and output pulse waveforms. The most significant difference among these types of fibers is modal dispersion. Because modal dispersion is described in detail in Chapter 3, the difference due to modal dispersion is only briefly explained here.

When an optical pulse is input into an MMF, the optical power of the pulse is generally distributed to a huge number of the modes of the fiber. Different modes travel at different propagation speeds along the fiber, which means that different modes launched at the same time reach the output end of the fiber at different times. Therefore, the input pulse broadens in time as it travels along the MMF. This pulse broadening effect, well known as *modal dispersion*, is significantly observed in SI MMFs. As shown in Figure 1.1, different rays travel along paths with different lengths; here each distinct ray can be thought of as a mode in a

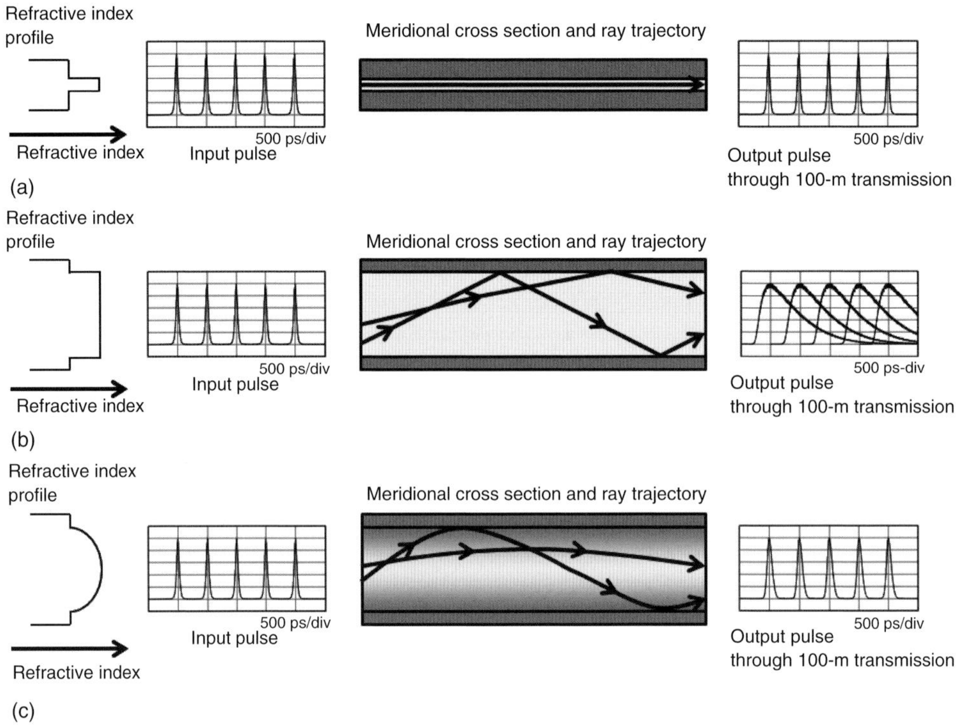

Refractive index profile
Meridional cross section and ray trajectory
500 ps/div
Refractive index
Input pulse
500 ps/div
Output pulse through 100-m transmission
(a)
Refractive index profile
Meridional cross section and ray trajectory
500 ps/div
Input pulse
500 ps-div
Output pulse through 100-m transmission
Refractive index
(b)
Refractive index profile
Meridional cross section and ray trajectory
500 ps/div
Input pulse
500 ps/div
Output pulse through 100-m transmission
Refractive index
(c)

Figure 1.1 Refractive index profiles and ray trajectories in (a) SI SMF, (b) SI, and (c) GI MMFs.

simple interpretation. The rays travel at the same velocity along their optical paths because of the constant refractive index throughout the core region in an SI MMF. Consequently, the same velocity and different path lengths result in different propagation speeds along the fiber, which causes a wide pulse spread in time. The pulse broadening caused by modal dispersion seriously limits the transmission capacity of MMFs because overlapping of the broadened pulses induces intersymbol interference and disrupts correct signal detection, thereby increasing the bit error rate (see Chapter 7) [7].

Modal dispersion is generally a dominant factor in pulse broadening in MMFs. However, modal dispersion can be dramatically reduced by forming a nearly parabolic refractive index profile in the core region of a GI MMF, which allows a much higher bandwidth and hence higher speed data transmission [8]. The optical ray is confined to near the core axis, corresponding to a lower order mode, and travels a shorter geometrical length at a slower light velocity along the path because of the higher refractive index. The sinusoidal ray passing through near the core–cladding boundary, which is considered as a higher order mode, travels a longer geometrical length at a faster velocity along the path, particularly in the lower refractive index region far from the core axis. As a result, the output times from the fiber end of rays through the shorter geometrical length at the

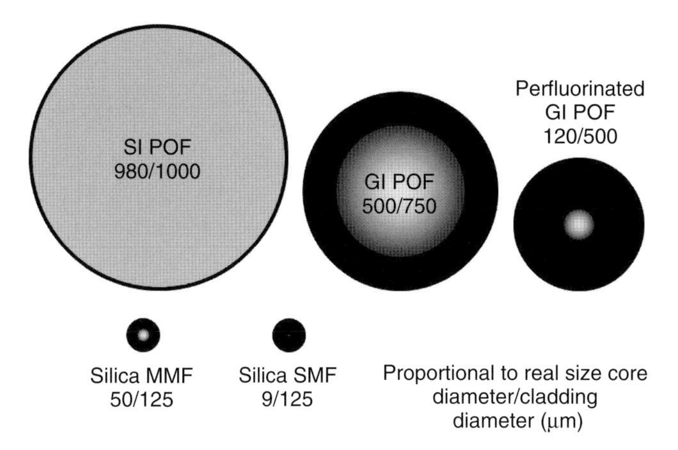

Figure 1.2 Cross sections of typical optical fibers.

slower velocity and of those through the longer geometrical length at the faster velocity can be almost the same because of the optimum refractive index profile. Therefore, GI MMF can realize high-speed data transmission. On the other hand, SMF has no modal dispersion in principle because only one mode is contained in it owing of its extremely small core. Thus, SMFs provide even higher bandwidth. Cross sections of typical optical fibers are shown in Figure 1.2.

Moreover, optical fibers can be categorized according to their base materials: silica glass and polymer. Silica SMFs are widely used in long-haul communication systems such as undersea networks because of the extremely low attenuation and high bandwidth [9]. On the other hand, optical fibers are currently required for data communication in LANs in homes, offices, hospitals, vehicles, and aircraft, as well as in long-haul telecommunication. As optical signal processing and transmission speeds have increased with developments in information and communication technologies, metal wiring has become a bottleneck for high-speed data transmission systems and large parallel processing computer systems. This is because electrical wiring causes significant problems, including electromagnetic interference, high signal reflection, high power consumption, and heat generation [10]. Thus, optical networking is expected to be used even in very short-reach networks. However, the connection of SMFs requires accurate alignment using expensive and precise connectors because of the extremely small core diameters of less than 10 μm, which is almost 1/10 of the diameter of a human hair. The connection requirements of SMFs results in high installation costs, especially in LANs where many connections are expected [11]. Silica MMFs have larger core diameters (50 or 62.5 μm) than SMFs. However, in addition to the modal noise and unstable bandwidth performance mentioned above, the core diameters of MMFs cannot be enlarged sufficiently for rough connections. Silica MMFs with larger core diameters are easily broken, because silica glass is inherently brittle. On the other hand, POFs can have much larger core diameters, from hundreds of micrometers to nearly 1000 μm. SI POFs with core diameters of almost 1000 μm are commercially available. The large core diameters of SI POFs enable rough and easy connections, so SI POFs are expected to be used as the transmission media in LANs. However, as the required data rate increases, SI POFs cannot achieve reliable data transmission because of the low bandwidths induced by the large modal dispersion (see Chapter 3). In contrast, GI POFs have been investigated as the transmission media in high-speed and short-reach networks [12, 13]. This is because GI POFs can realize stable and reliable high-speed communication owing to the high bandwidth, and because GI POFs can have much larger core diameters owing to the inherent flexibility of polymers. Therefore, GI POFs allow rough connection and easy handling [14–19], which dramatically reduces the installation cost of networks, particularly LANs [20, 21]. Thus, GI POFs are attracting a great deal of attention in consumer use because of their user-friendly characteristics. Although several types of POFs, such as single mode, SI, multistep index, multicore, GI, and microstructured, have been reported [14, 22], this book explains mainly representative SI and GI POFs.

1.2 Plastic Optical Fiber

The peripheral component of communication networks, referred to as *the last mile*, is estimated to account for ~95% of the overall network. Electrical wiring such as unshielded twisted pair (UTP) and coaxial cables has most often been adopted in LANs. However, its bandwidth and transmission distance are severely limited, and it is difficult to realize high-speed data transmission on the order of gigabits per second at distances greater than 100 m using UTP. On the other hand, silica-based SMFs used in backbone systems can achieve extremely high data rates and long-distance communication. However, precise and time-consuming techniques are demanded for termination, connection, and branching because of their very small diameters of less than 10 μm, which induces high cost of installation of LAN systems with huge numbers of connections and junctions. In contrast, POFs, which consist of a polymer core and cladding, can have much larger core diameters (up to 1000 μm) than silica-based optical fibers because of their inherent flexibility, although POFs exhibit relatively high attenuation. POFs cannot penetrate human skin or be broken by bending and physical impact. Highly accurate alignment is not required for POF connections because of the large core. These characteristics enable easy and low-cost installation and safe handling.

The demand for high-speed communication over private intranets and the Internet is growing explosively as the available data volume in personal devices increases. In particular, a strong demand for more realistic video images, such as 8 K and/or 3D, and more realistic face-to-face communication requires higher resolution, more natural color, and higher frame rates. Thus, increasingly large bit rates for data transmission are required in high-resolution displays and cameras. Consequently, POFs are attracting a great deal of attention because electrical wiring causes critical problems as described above [11]. The commonly cited advantages of optical fibers are their remarkably high bandwidth, immunity to electromagnetic interference, and immunity to crosstalk. With the demand for high-speed data processing and communication systems, GI POFs have become promising candidates for optical interconnects as well as optical networking in LANs because of their high bandwidth, in addition to their advantages for consumer use such as high tolerance to misalignment and bending, high mechanical strength, and long-term reliability [13, 16, 18, 19, 23].

The first SI POF was reported by DuPont in the mid-1960s, around the same time as the invention of the silica fiber. The SI POF was first commercialized by Mitsubishi Rayon in 1975. Asahi Chemical and Toray also entered the market. However, early POFs had quite high attenuation (for example, 1000 dB/km), and the transmission length was critically limited to only several meters. Thus, limited applications were considered, such as light guiding, illumination, and sensors, rather than data transmission. Analyses of attenuation in POFs clarified that the high attenuation was caused mainly by extrinsic factors such as contamination and imperfections introduced during fiber fabrication and were not intrinsic and inevitable (see Chapter 2). Indeed, POFs exhibited low attenuation near the theoretical prediction when the monomer was purified and contaminants were eliminated [24]. Thus, the attenuation of the SI POF became low enough for applications in premise networks. The achievements of low-loss POFs are shown in Figure 1.3.

Although the attenuation in SI POFs was dramatically reduced, their bandwidth, another important parameter, is severely limited by modal dispersion and is far from the requirements for high-speed data transmission. On the other hand, the first GI POF was reported in 1982 [25]. The GI POF theoretically has a high bandwidth (see Chapter 3); however, the attenuation was greater than 1000 dB/km. Thus, the development of POFs again encountered attenuation problems. The first GI POF was fabricated by copolymerization of more than two monomers with different refractive indices, and a graded refractive index profile was formed by controlling the composition distribution of the copolymer according to the

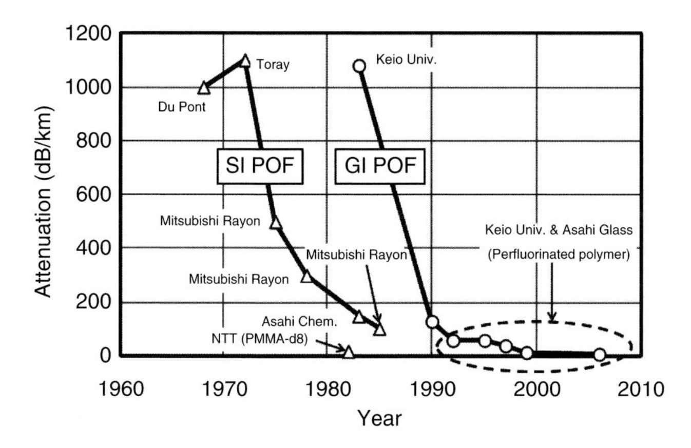

Figure 1.3 Reduction in attenuation of POFs.

monomer reactivity ratios. Impurities and contaminants were suspected to be a dominant factor in the high attenuation. However, microscopic heterogeneous structures caused by the composition distribution of the copolymer were found to cause high attenuation in GI POFs. Therefore, a GI POF with a low-molecular-weight dopant was invented in 1991 [12, 20]. The dopant had a higher refractive index than the base polymer of the GI POF, and a refractive index profile corresponding to the dopant concentration distribution was obtained in this GI POF by interfacial gel polymerization (see Chapter 5). The formation of GI POFs requires a dopant that has a higher refractive index than the host polymer. The addition of the dopant countered the trend toward purification to reduce attenuation because the dopant was considered a type of impurity. However, a low-loss, high-bandwidth GI POF with the dopant was developed. This was a breakthrough for high-speed POF networks. The attenuation was further reduced, and the bandwidth was further enhanced by fluorination of the polymer [26, 27]. After various reports of new high-bandwidth records, the GI POF realized 40 Gb/s data transmission over 100 m [28]. The achievements of high-speed data transmission in POF networks are plotted in Figure 1.4. The bit rate–distance product (vertical axis) indicates the data transmission performance, and higher values indicate that a higher bit rate and/or longer distance is available in the optical link. The transmission performance of the SI POF (diamonds) was limited to hundreds of megabits per second over 100 m because of the large modal dispersion. The bit rate of poly(methyl methacrylate)-based GI POFs (squares) increased to several gigabits per second over 100 m because of the graded refractive index profile. Both the transmission distance and bit rate of perfluorinated-polymer-based GI POFs (triangles) were dramatically improved because of the low attenuation and low material dispersion inherent to perfluorinated polymers (see Chapters 2 and 3). The bit rate of the perfluorinated GI POFs fabricated by coextrusion (circles) reached 40 Gb/s over 100 m, a breakthrough achievement.

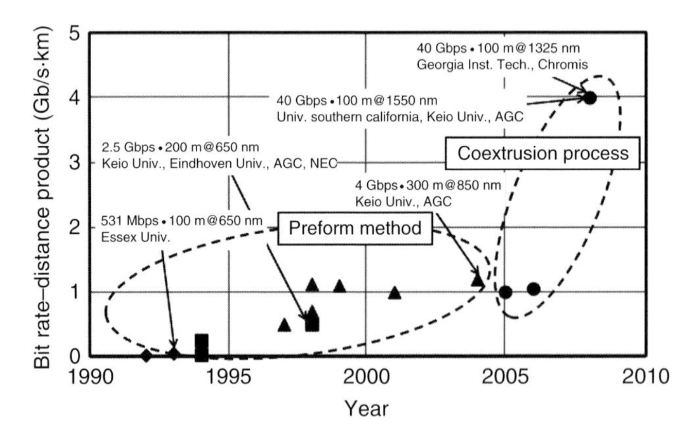

Figure 1.4 Enhancement in bit rate–distance product of POFs. ♦: PMMA (poly(methyl methacrylate))-based SI POF, ■: PMMA-based GI POF by preform method, ▲: perfluorinated GI POF by preform method, and •: perfluorinated GI POF by coextrusion process.

References

1. Hecht, J. (2004) *City of Light: The Story of Fiber Optics*, Oxford University Press, New York.
2. Kao, K. and Hockham, G.A. (1966) Dielectric-fibre surface waveguides for optical frequencies. *Proc. IEE*, **113** (7), 1151–1158.
3. Kapron, F., Keck, D.B., and Maurer, R.D. (1970) Radiation losses in glass optical waveguides. *Appl. Phys. Lett.*, **17** (10), 423–425.
4. Born, M. and Wolf, E. (1999) *Principles of Optics: Electromagnetic Theory of Propagation, Interference and Diffraction of Light*, Cambridge University Press.
5. Hecht, E. (2002) *Optics*, 4th edn, Addison-Wesley.
6. Hecht, J. and Long, L. (2002) *Understanding Fiber Optics*, Prentice Hall, Columbus, OH.
7. Nowell, M.C., Cunningham, D.G., Hanson, D.C., and Kazovsky, L.G. (2000) Evaluation of Gb/s laser based fibre LAN links: review of the Gigabit Ethernet model. *Opt. Quantum Electron.*, **32** (2), 169–192.
8. Gloge, D. and Marcatili, E.A.J. (1973) Multimode theory of graded-core fibers. *Bell Syst. Tech. J.*, **52** (9), 1563–1578.
9. Olshansky, R. (1979) Propagation in glass optical waveguides. *Rev. Mod. Phys.*, **51** (2), 341–367.
10. Ball, P. (2012) Computer engineering: feeling the heat. *Nature*, **492** (7428), 174–176.
11. Polishuk, P. (2006) Plastic optical fibers branch out. *IEEE Commun. Mag.*, **44** (9), 140–148.
12. Koike, Y. (1991) Optical resin materials with distributed refractive index, process for producing the materials, and optical conductors using the materials. US Patent 5541247, JP Patent 3332922, EU Patent 0566744, KR Patent 170358, CA Patent 2098604, originally filed in 1991.
13. Koike, Y. (1991) High-bandwidth graded-index polymer optical fibre. *Polymer*, **32** (10), 1737–1745.
14. Zubia, J. and Arrue, J. (2001) Plastic optical fibers: an introduction to their technological processes and applications. *Opt. Fiber Technol.*, **7** (2), 101–140.
15. Ishigure, T., Hirai, M., Sato, M., and Koike, Y. (2004) Graded-index plastic optical fiber with high mechanical properties enabling easy network installations. *J. Appl. Polym. Sci.*, **91** (1), 404–416.
16. Makino, K., Ishigure, T., and Koike, Y. (2006) Waveguide parameter design of graded-index plastic optical fibers for bending-loss reduction. *J. Lightwave Technol.*, **24** (5), 2108–2114.
17. Makino, K., Kado, T., Inoue, A., and Koike, Y. (2012) Low loss graded index polymer optical fiber with high stability under damp heat conditions. *Opt. Express*, **20** (12), 12893–12898.
18. Makino, K., Akimoto, Y., Koike, K., Kondo, A., Inoue, A., and Koike, Y. (2013) Low loss and high bandwidth polystyrene-based graded index polymer optical fiber. *J. Lightwave Technol.*, **31** (14), 2407.
19. Makino, K., Nakamura, T., Ishigure, T., and Koike, Y. (2005) Analysis of graded-index polymer optical fiber link performance under fiber bending. *J. Lightwave Technol.*, **23** (6), 2062–2072.
20. Koike, Y., Ishigure, T., and Nihei, E. (1995) High-bandwidth graded-index polymer optical fiber. *J. Lightwave Technol.*, **13** (7), 1475–1489.
21. Koike, Y. and Ishigure, T. (2006) High-bandwidth plastic optical fiber for fiber to the display. *J. Lightwave Technol.*, **24** (12), 4541–4553.
22. van Eijkelenborg, M.A., Large, M.C.J., Argyros, A., Zagari, J., Manos, S., Issa, N.A., Bassett, I., Fleming, S., McPhedran, R.C., de Sterke, C.M., and Nicorovici, N.A.P. (2001) Microstructured polymer optical fibre. *Opt. Express*, **9** (7), 319–327.
23. Koike, Y. and Koike, K. (2011) Polymer optical fibers, in *Encyclopedia of Polymer Science and Technology*, John Wiley & Sons, Inc.
24. Kaino, T., Fujiki, M., and Jinguji, K. (1984) Preparation of plastic optical fibers. *Rev. Electr. Commun. Lab.*, **32** (3), 478–488.
25. Koike, Y., Kimoto, Y., and Ohtsuka, Y. (1982) Studies on the light-focusing plastic rod. 12: the GRIN fiber lens of

methyl methacrylate-vinyl phenylacetate copolymer. *Appl. Opt.*, **21** (6), 1057–1062.

26. Koike, Y. and Naritomi, M. (1994) Graded-refractive-index optical plastic material and method for its production. JP Patent 3719733, US Patent 5783636, EU Patent 0710855, KR Patent 375581, CN Patent L951903152, TW Patent 090942, originally filed in 1994.

27. Naritomi, M., Murofushi, H., and Nakashima, N. (2004) Dopants for a perfluorinated graded index polymer optical fiber. *Bull. Chem. Soc. Jpn.*, **77** (11), 2121–2127.

28. Polley, A. and Ralph, S.E. (2007) Mode coupling in plastic optical fiber enables 40-Gb/s performance. *IEEE Photonics Technol. Lett.*, **19** (16), 1254–1256.

2
Transmission Loss

It would not be an exaggeration to say that the history of plastic optical fibers (POFs) has been a history of the attempts to reduce their transmission loss. The transmission loss limits how far a signal can propagate in the fiber before the optical power becomes too weak to be detected. It measures the amount of light lost between the input and output; it is normally expressed in decibels and defined as

$$dB = -10 \log_{10}\left(\frac{P_{out}}{P_{in}}\right). \tag{2.1}$$

This is the sum of all the losses. The various mechanisms contributing to the losses in POFs are essentially similar to those for glass optical fibers (GOFs), but the relative magnitudes are different. Figure 2.1 shows the loss factors for POFs, which are divided into intrinsic and extrinsic factors. Although extrinsic factors such as contaminants or waveguide imperfections can sometimes cause important losses, once an optimum fabrication process has been achieved, they can be ignored. The intrinsic factors are further classified into absorption and scattering losses. This chapter is devoted to a discussion of these influences in POFs and how the transmission loss has been reduced.

2.1
Absorption Loss

2.1.1
Electronic Transition Absorption

The absorption of light in POF materials depends on its frequency or wavelength because materials have various energy levels that are involved in absorption transitions. In the light wavelengths used for data communication with POFs, the intrinsic absorption losses are caused by electronic transition absorptions and/or molecular vibration absorptions. The electronic transition absorption peaks typically appear at ultraviolet wavelengths, and their absorption tails influence the transmission losses of POFs. For example, a POF with a poly(methyl methacrylate) (PMMA) core exhibits $n-\pi^*$ transitions due to the ester groups in methyl methacrylate (MMA) molecules, $n-\sigma^*$ transitions of S–H bonds in chain-transfer

Fundamentals of Plastic Optical Fibers, First Edition. Yasuhiro Koike.

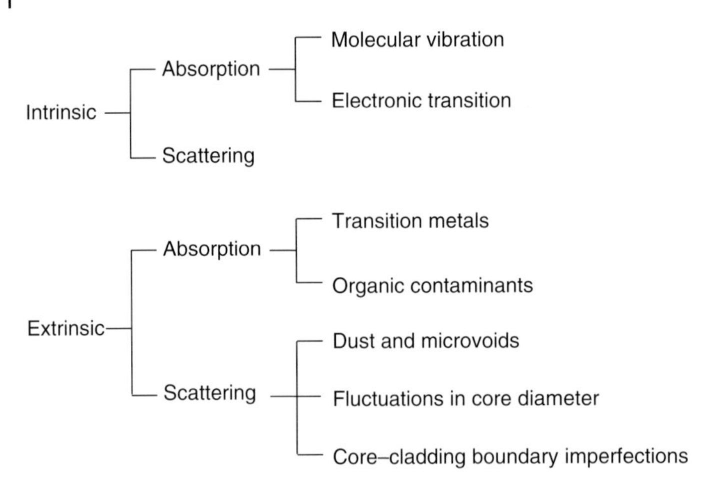

Figure 2.1 Classification of intrinsic and extrinsic factors affecting POF attenuation.

agents, and $\pi-\pi^*$ transitions of azo groups when azo compounds are used as an initiator for polymerization. The most significant absorption is the transition of the $\mathrm{n}-\pi^*$ orbital of the double bond within the ester group. According to Urbach's rule [1], the electronic transition absorption spectra have exponential tails, given by

$$\alpha_\mathrm{e} = A \exp\left(\frac{B}{\lambda}\right). \tag{2.2}$$

Here α_e (dB/km) is the electronic transition absorption loss, and λ (nm) is the incident light wavelength. For PMMA, the substance-specific constants A and B have been identified as 1.58×10^{-12} and 1.15×10^4, respectively [2]. Hence, the value of α_e for PMMA is less than 1 dB/km at 500 nm. On the other hand, polystyrene (PSt) [2] and polycarbonates (PCs) [3], which are also widely used for POFs, exhibit considerably larger absorptions losses, as shown in Figure 2.2. This is due to the $\pi-\pi^*$ transition of phenyl groups included in PSt and PC. The bandgap energy between the π and π^* levels is comparable to the photon energies of visible light, and the tails are dramatically shifted to longer wavelengths as the conjugation length increases.

2.1.2 Molecular Vibration Absorption

Molecular vibration absorptions are typically observed at infrared wavelengths, which correspond to the resonance frequencies for fundamental molecular vibrations. Because of the molecular potential anharmonicity, however, overtone and combination absorption bands also appear at visible and near-infrared wavelengths. The predominant factor for attenuation in POFs has been the stretching overtone absorptions of C–H bonds.

To understand the overtone absorption of POF materials, let us consider individual chemical bonds in a polymer as the equivalent diatomic molecules with only

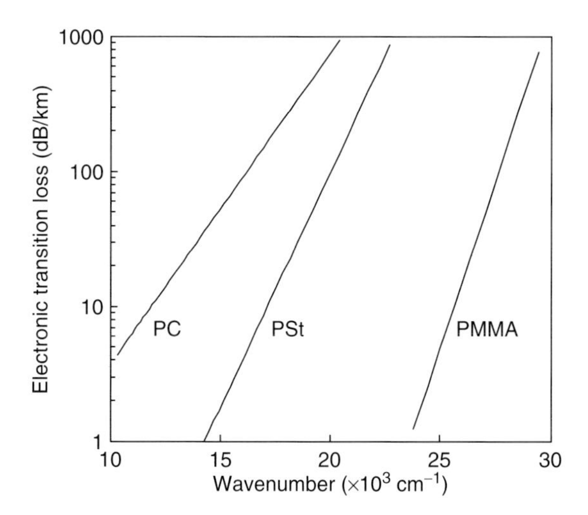

Figure 2.2 Electronic transition loss of PMMA, PSt, and PC.

one degree of vibrational freedom (stretching vibration). The anharmonic potential curve of the diatomic molecule is often approximated by the Morse function as [4]

$$V(r) = hcD_e[1 - e^{-a_M(r-r_e)}]^2. \tag{2.3}$$

Here r, r_e, and D_e are the atomic distance, equilibrium bond length, and potential-well depth, respectively. Further, h is Planck's constant and c is the light velocity in vacuum. The potential-curve shapes depend on the parameters a_M and D_e. By analytically solving the Schrödinger equation with the anharmonic potential (Equation 2.3), one can obtain the vibrational energy levels of the Morse molecular oscillator:

$$E_\upsilon = \left(\upsilon + \frac{1}{2}\right)h\nu_e - \left(\upsilon + \frac{1}{2}\right)^2 h\nu_e\chi_e \quad (\upsilon = 0, 1, 2, \ldots \upsilon_{max}). \tag{2.4}$$

The first term on the right-hand side corresponds to the energy levels in the harmonic oscillator approximation for small displacements from equilibrium, where $V(r) \approx (1/2)k_e(r - r_e)^2$. The harmonic oscillator frequency ν_e is given by

$$\nu_e = \frac{1}{2\pi}\left(\frac{k_e}{m_e}\right)^{\frac{1}{2}} \quad \text{with} \quad k_e = 2hcD_e a_M^2. \tag{2.5}$$

Here m_e is the effective mass defined in terms of the atom masses m_1 and m_2 as $m_e = m_1m_2/(m_1 + m_2)$. On the other hand, the second additional term in Equation 2.4 is the contribution of the oscillator anharmonicity, which depends on the anharmonicity constant

$$\chi_e = \frac{\nu_e}{4cD_e} = \frac{a_M}{4\pi}\left(\frac{h}{2m_e cD_e}\right)^{\frac{1}{2}}. \tag{2.6}$$

As shown in Figure 2.3, the anharmonicity results in convergence of the energy level at high excitation because the additional term becomes more important for

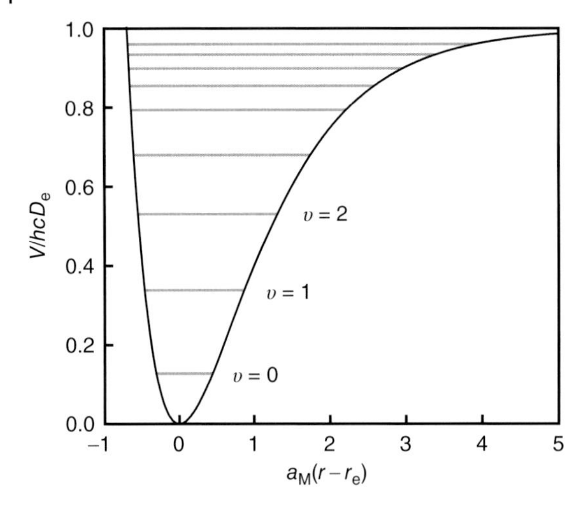

Figure 2.3 Morse potential curve and vibrational energy levels.

higher υ values in Equation 2.4. Consequently, the molecule has the finite energy level number υ_{max}, where $\upsilon_{max} < 1/(2\chi_e) - 1/2$. This corresponds to the fact that the molecule can be dissociated. The dissociation energy, or bonding energy, of the Morse diatomic molecule is given by

$$D_0 = D_e - \frac{E_0}{hc}, \tag{2.7}$$

where E_0 is the zero-point energy for the ground state.

Molecular vibration absorption is the vibrational transition from the ground state to an excited state. From Equation 2.4, the resonance absorption frequency for the transition to the υth excited state can be expressed by

$$\nu_\upsilon = \upsilon\nu_e - \upsilon(\upsilon+1)\nu_e\chi_e \quad (\upsilon \geq 1). \tag{2.8}$$

Here, ν_υ is the overtone absorption frequency, except for $\upsilon = 1$ which is the fundamental vibration absorption. By substituting $\nu_e = \nu_1/(1-2\chi_e)$ in Equation 2.8, we can obtain the following useful expression for the overtone absorption frequency:

$$\nu_\upsilon = \frac{\upsilon\nu_1 - \upsilon(\upsilon+1)\nu_1\chi_e}{1-2\chi_e} \quad (\upsilon \geq 2). \tag{2.9}$$

All the overtone absorption frequencies can be calculated using this equation if the anharmonicity constant χ_e and the fundamental vibration frequency ν_1 are known. On the other hand, the values of ν_e and $\nu_e\chi_e$ are also often estimated to evaluate the bond potential by using the measured overtone absorption frequencies in Equation 2.8 [5–7].

The overtone absorption band intensity can be evaluated on the basis of quantum chemical analyses of the transition moment, which determines the absorption transition rate. According to Mecke's approach [8–10], however, we can roughly

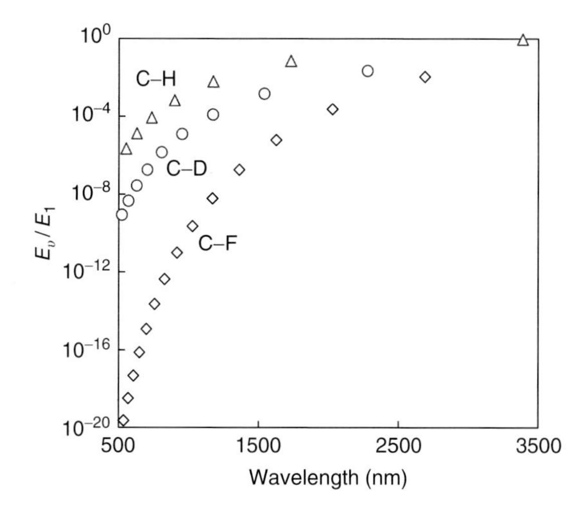

Figure 2.4 Calculated spectral overtone positions and normalized integral band strengths for different C–X vibrations.

estimate the intensity with the following analytical expression for the υth overtone absorption band intensity:

$$E_\upsilon = \frac{f_1(\chi_e)}{f_\upsilon(\chi_e)} E_1 \quad (\upsilon \geq 2) \tag{2.10}$$

with

$$f_\upsilon(\chi_e) = \frac{1}{\upsilon(\chi_e^{-1} - 2\upsilon - 1)} \frac{\Gamma(\chi_e^{-1} - 1)}{\Gamma(\chi_e^{-1} - \upsilon - 1)\Gamma(\upsilon + 1)}. \tag{2.11}$$

Here, E_1 is the intensity of the fundamental vibration absorption band. Note that Equation 2.10 can be derived by assuming a smooth dipole moment curve, by which bonds have no strong vibrational coupling with the other bonds in the molecule. Figure 2.4 shows the relative integral band intensities E_υ/E_1 for different C–X bonds as a function of the absorption peak wavelengths, which were calculated using the reported values for χ_e, υ_1, and E_1 in Equations 2.9 and 2.10 [9]. In the visible to near-infrared region, the C–D and C–F overtone absorption intensities are several orders of magnitude lower than those of C–H bonds because of the much higher overtone orders. This implies that fiber attenuation can be significantly reduced by replacing the hydrogen atoms with atoms such as deuterium and fluorine.

2.1.3 Effect of Fluorination on Attenuation Spectra of POFs

As mentioned in the previous subsection, absorption losses in POFs have been attributed mainly to the stretching overtone absorption of the C–H bonds, resulting in much higher attenuation in POFs than in GOFs. Therefore, the

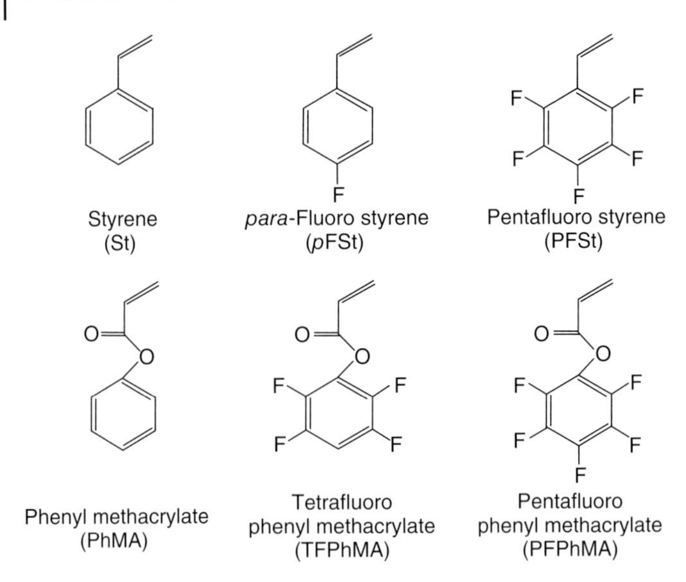

Figure 2.5 Chemical structures of styrene (St), phenyl methacrylate (PhMA), and their fluorinated counterparts.

attenuation in POFs has been reduced by replacing the hydrogen atoms in the fiber core base materials with heavier atoms such as deuterium, fluorine, and chloride. Now, perfluorinated (PF) POFs allow light transmission with little vibrational absorption, although they have limited applications because of their high material cost. Partially fluorinated polymers have been studied recently as low-cost POF materials with sufficiently low attenuation for various applications [11]. Fluorination was intended to reduce the C–H bond number density and thus the C–H overtone absorption. On the other hand, fluorination also shifts the peak wavelength of the aromatic C–H overtone absorption in polymers with partially fluorinated phenyl groups [12, 13]. This suggests that fluorination can affect the low-loss optical windows in partially fluorinated POFs. Nevertheless, attenuation in POFs has been analyzed by assuming that the C–H overtone absorption does not depend on the polymer molecular structures. Here, we introduce some effects of fluorination [14] on the C–H overtone absorption spectra of partially fluorinated materials, as shown in Figure 2.5.

Figure 2.6 shows the absorption spectra of PSt, poly(*p*FSt) (P*p*FSt), and poly(PFSt) (PPFSt) bulks with different fluorine contents in the phenyl groups (Figure 2.5), where the contributions of light scattering and electronic transition absorption are negligibly small. They have mainly third and fourth C–H overtone absorption bands, although some combination bands are also observed in the spectra of P*p*FSt and PPFSt. Note that PSt and P*p*FSt have both the aromatic and the aliphatic C–H absorption bands, whereas PPFSt has only the aliphatic band. The aromatic C–H absorption peak wavelengths of P*p*FSt are shorter than those of PSt, which are 1140 and 870 nm for the third and

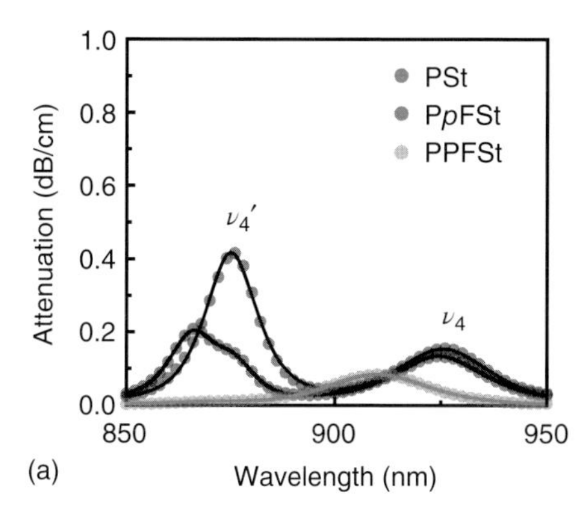

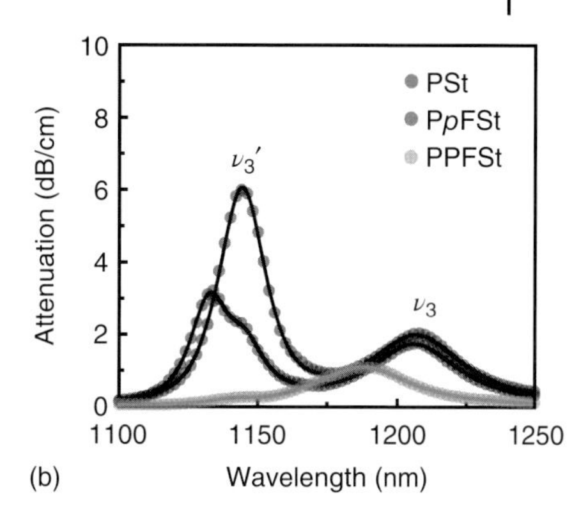

Figure 2.6 Attenuation spectra of PSt, P*p*FSt, and PPFSt at wavelengths of (a) 850–950 nm and (b) 1050–1250 nm. Closed circles are experimental data; solid lines are fitted theoretical curves. ν_x (ν_x') shows *x*th overtone vibration of aliphatic (aromatic) C–H bonds in PSt.

fourth overtone absorption bands, respectively. This is attributed to the electron withdrawing effect of fluorine [13], which changes the potential curve of the aromatic C–H bonds. Moreover, similar wavelength shifts are observed for the aliphatic C–H absorption bands in the spectra of PPFSt, whereas PSt and P*p*FSt have comparable peak wavelengths of the third and fourth overtones around 1200 and 930 nm, respectively. This suggests that the effect of the fluorine depends on the fluorinated position, fluorine content, and molecular structure.

Figure 2.7 shows the corresponding overtone absorption spectra of PPhMA, poly(TFPhMA) (PTFPhMA), and poly(PFPhMA) (PPFPhMA) bulks with different fluorine contents in the phenyl groups (Figure 2.5). They exhibit both

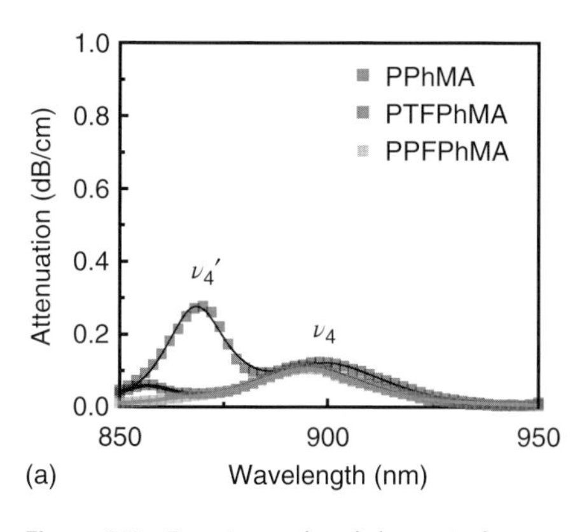

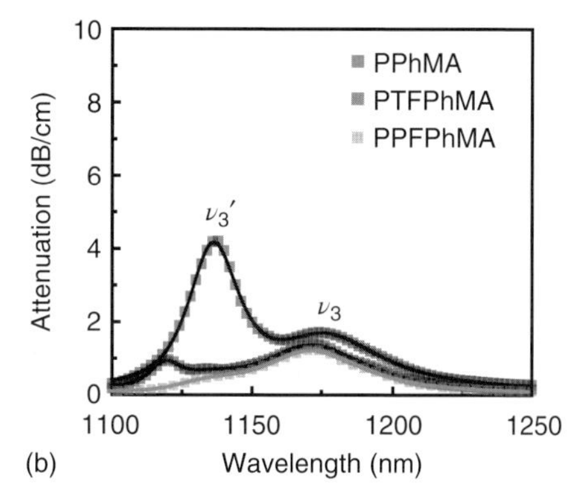

Figure 2.7 Experimental and theoretical attenuation spectra of PPhMA, PTFPhMA, and PPFPhMA at wavelengths of (a) 850–950 nm and (b) 1050–1250 nm. Closed squares are experimental data; solid lines are fitted theoretical curves. ν_x (ν_x') shows *x*th overtone vibration of aliphatic (aromatic) C–H bonds in PPhMA.

Table 2.1 Overtone absorption intensities of aromatic and aliphatic C–H bonds in PSt, PPhMA, and their fluorinated counterparts.

Polymer	Integral bandstrength (10^3 cm/mol)			
	ν'_3	ν'_4	ν_3	ν_4
PSt	2.60	0.24	2.92	0.20
P*p*FSt	1.33	0.14	2.64	0.21
PPFSt	—	—	1.94	0.15
PPhMA	2.45	0.23	2.06	0.16
PTFPhMA	1.54	0.20	1.94	0.19
PPFPhMA	—	—	1.86	0.19

the aromatic and aliphatic C–H overtone absorption, except for PPFPhMA which lacks aromatic C–H bonds. As is the case in P*p*FSt, fluorination shortens the peak wavelengths of the aromatic C–H absorption bands of PTFPhMA compared to those of PPhMA, which are 1130 and 870 nm for the third and fourth overtone absorption bands, respectively. However, the aliphatic C–H absorption bands are affected little by fluorination, resulting in comparable peak wavelengths of the third and fourth overtone bands around 1175 and 900 nm, respectively. The different effects of fluorine compared to those in fluorinated PSts are related to the molecular structures of PPhMA materials, in which the carbonyl group appears between the aromatic and aliphatic C–H bonds. For PPhMA materials, the carbonyl group can disturb the mutual interaction of the aromatic and aliphatic C–H bonds, whereas fluorination of the aromatic C–H bonds can directly affect the potential curves of the aliphatic C–H bonds in fluorinated PSts.

Fluorination can also affect the absorption band intensities of C–H bonds by changing the potential curve or anharmonicity constant, as shown in Equation 2.10. Table 2.1 shows the averaged absorption band intensities of the aromatic and aliphatic C–H bonds in PSt, PPhMA, and their fluorinated counterparts. They were estimated by fitting the absorption spectra to the sum of the pseudo-Voigt profiles with the band intensities, bandwidths, and resonance frequencies [15, 16]. The results show that fluorination decreases the absorption band intensities of the aromatic and/or aliphatic C–H bonds in fluorinated PSts. On the other hand, the aliphatic C–H absorption intensities are little affected by fluorination in fluorinated PPhMA. The intensity reduction tendencies correspond to the absorption peak wavelength shifts, as shown in Figures 2.6 and 2.7. These results suggest that the attenuation reduction efficiency of fluorination also depends on the substitution position in the benzene ring and the chemical structure.

2.2 Scattering Loss

2.2.1 Definition of Scattering Loss

Scattering losses in polymers arise from microscopic variations in the material density. When natural light of intensity I_0 passes through a distance y, and its intensity is reduced to I by the scattering loss, the turbidity τ is defined as

$$\frac{I}{I_0} = \exp(-\tau y). \tag{2.12}$$

Because τ corresponds to the summation of light scattered in all directions, it is given as

$$\tau = \pi \int_0^{\pi} (V_V + V_H + H_V + H_H) \sin\theta \, d\theta. \tag{2.13}$$

Here, V and H denote vertical and horizontal polarization, respectively. The symbol A and subscript B in the expression for a scattering component A_B represent the directions of the polarizing phase of scattered light and incident light, respectively. θ is the scattering angle in relation to the direction of the incident light. In structureless liquids or randomly oriented bulk polymers, these intensities are given by the following equations:

$$H_V = V_H, \tag{2.14}$$

$$H_H = V_V \cos^2\theta + H_V \sin^2\theta. \tag{2.15}$$

Here, the isotropic part V_V^{iso} of V_V is given as follows:

$$V_V^{iso} = V_V - \frac{4}{3} H_V. \tag{2.16}$$

By substituting Equations 2.14–2.16 into Equation 2.13, τ can be rewritten as follows:

$$\tau = \pi \int_0^{\pi} \left\{ \left(1 + \cos^2\theta\right) V_V^{iso} + \frac{(13 + \cos^2\theta)}{3} H_V \right\} \sin\theta \; d\theta. \tag{2.17}$$

Furthermore, the intensity of the isotropic light scattering, V_V^{iso}, and the anisotropic light scattering, H_V, can be expressed by Equation 2.18 [17] and Equation 2.19 [18], respectively, as

$$V_V^{iso} = \frac{\pi^2}{9\lambda_0^4}(n^2 - 1)^2 (n^2 + 2)^2 k_B T \beta, \tag{2.18}$$

$$H_V = \frac{16\pi^4}{135\lambda_0{}^4}(n^2 + 2)^2 N \langle \delta^2 \rangle. \tag{2.19}$$

Here, λ_0 is the wavelength of light in vacuum, n is the refractive index, k_B is the Boltzmann constant, T is the absolute temperature, β is the isothermal compressibility, N is the number of scattering units per unit volume, and $\langle\delta^2\rangle$ is the mean square of the anisotropic parameter of polarizability per scattering unit. Finally, from the definition of the turbidity τ in Equation 2.12, the light scattering loss α_s (dB/km) is related to the turbidity τ (cm^{-1}) by

$$\alpha_s = 4.342 \times 10^5 \tau. \tag{2.20}$$

2.2.2 Heterogeneous Structure and Excess Scattering

On the other hand, when polymers have large heterogeneities in their higher order structures, they exhibit considerably stronger light scattering than the theoretical values calculated by the above equations. Scattered light interferes with itself, and the intensity of isotropic scattering V_V^{iso} shows an angular dependence. V_V^{iso} is separated into two terms as follows:

$$V_V^{\text{iso}} = V_{V1}^{\text{iso}} + V_{V2}^{\text{iso}}. \tag{2.21}$$

Here, V_{V1}^{iso} denotes a background intensity that is independent of the scattering angle, whereas V_{V2}^{iso} is the excess scattering with an angular dependence due to large heterogeneities. By substituting Equation 2.21 into Equation 2.17, τ can be written as follows:

$$\tau = \pi \int_0^{\pi} \left\{ \left(1 + \cos^2\theta\right) \left(V_{V1}^{\text{iso}} + V_{V2}^{\text{iso}}\right) + \frac{(13 + \cos^2\theta)}{3} H_V \right\} \sin\theta \; d\theta. \tag{2.22}$$

For V_{V2}^{iso}, Debye and Bueche derived [19]

$$V_{V2}^{\text{iso}} = \frac{4\langle\eta^2\rangle\pi^3}{\lambda_0^4} \int_0^{\infty} \frac{\sin(ksr)}{ksr} r^2 \gamma(r)\, dr, \tag{2.23}$$

where $\langle\eta^2\rangle$ denotes the mean-squared average of the fluctuations of all the dielectric constants, $k = 2\pi/\lambda$, and $s = 2\ \sin(\theta/2)$. Further, λ and λ_0 are the wavelengths of light in a specimen and under vacuum, respectively; $\gamma(r)$ refers to the correlation function defined by $\eta_i\eta_j/\langle\eta^2\rangle$, where η_i and η_j are the fluctuations of the dielectric constant at positions i and j, respectively, from the average. Assuming that the correlation function is expressed by

$$\gamma(r) = \exp(-r/D), \tag{2.24}$$

Equation 2.23 is simply integrated to give

$$V_{V2}^{\text{iso}} = \frac{8\pi^3\langle\eta^2\rangle D^3}{\lambda_0^4(1 + k^2s^2D^2)^2}. \tag{2.25}$$

Here, D is called the *correlation length* and is a measure of the size of the heterogeneous structure inside the bulk. The turbidity τ is divided into three terms, namely,

$$\tau = \tau_1^{\text{iso}} + \tau_2^{\text{iso}} + \tau^{\text{aniso}}, \tag{2.26}$$

where τ_1^{iso} is the turbidity from $V_{\text{V1}}^{\text{iso}}$ scattering, τ_2^{iso} is that from $V_{\text{V2}}^{\text{iso}}$, and τ^{aniso} is that from the anisotropic scattering H_{V}. From Equation 2.22, these terms are given as follows:

$$\tau_1^{\text{iso}} = \pi \int_0^{\pi} (1 + \cos^2\theta) V_{\text{V1}}^{\text{iso}} \sin\theta \; d\theta = \frac{8}{3}\pi V_{\text{V1}}^{\text{iso}}, \tag{2.27}$$

$$\begin{aligned} \tau_2^{\text{iso}} &= \pi \int_0^{\pi} (1 + \cos^2\theta) V_{\text{V2}}^{\text{iso}} \sin\theta \; d\theta \\ &= \frac{32D^3\langle\eta^2\rangle\pi^4}{\lambda_0{}^4} \left\{ \frac{(b+2)^2}{b^2(b+1)} - \frac{2(b+2)}{b^3} \ln(b+1) \right\}, \end{aligned} \tag{2.28}$$

$$b = 4k^2D^2, \tag{2.29}$$

$$\tau^{\text{aniso}} = \pi \int_0^{\pi} \frac{(13 + \cos^2\theta)}{3} H_{\text{V}} \sin\theta d\theta = \frac{80}{9}\pi H_{\text{V}}. \tag{2.30}$$

The light scattering loss α (dB/km) is obtained by substituting Equation 2.26 into Equation 2.20. The losses corresponding to each turbidity in Equations 2.27–2.30 are defined as α_{s}, α_1^{iso}, α_2^{iso}, and α^{aniso}, respectively; that is,

$$\alpha_{\text{s}} = \alpha_1^{\text{iso}} + \alpha_2^{\text{iso}} + \alpha^{\text{aniso}}. \tag{2.31}$$

Experimentally, $V_{\text{V1}}^{\text{iso}}$ and $V_{\text{V2}}^{\text{iso}}$ are separated as follows: By rearranging Equation 2.25, a plot of $V_{\text{V2}}^{\text{iso}-1/2}$ versus s^2 (the Debye plot) yields a straight line, and the correlation length D can be determined by $D = (\lambda/2\pi)/(\text{slope/intercept})^{1/2}$. Therefore, by gradually changing the $V_{\text{V2}}^{\text{iso}}$ value from 0 to the observed V_{V}, the $V_{\text{V2}}^{\text{iso}}$ at which the Debye plot becomes closest to a straight line is obtained by a least-squares technique.

2.2.3
Origin of Excess Scattering in PMMA

Light scattering in amorphous optical polymers such as PMMA, PSt, and PC has been studied extensively to investigate local structures ranging in size from hundreds to thousands of angstroms. In particular, PMMA has received considerable attention because of its application in POFs and optical waveguides. Even in highly purified PMMA glasses in which no anisotropic domains such as bundles of parallel molecular chains or folded chains were observed by neutron or X-ray scattering, large heterogeneities with dimensions of approximately 1000 Å were invariably observed in V_{V} scattering. Consequently, the scattering losses were as large as several hundred decibels per kilometer. According to Einstein's fluctuation theory, however, the intensity of the isotropic $V_{\text{V}}^{\text{iso}}$ should be given by Equation 2.18. Using the published data of $\beta = 3.55 \times 10^{-11}$ cm^2/dyn around the glass transition temperature T_{g} for PMMA bulk [20] and assuming freezing conditions, $V_{\text{V}}^{\text{iso}}$ at room temperature and a wavelength of 633 nm is 2.61×10^{-6} cm^{-1}. By using Equations 2.17

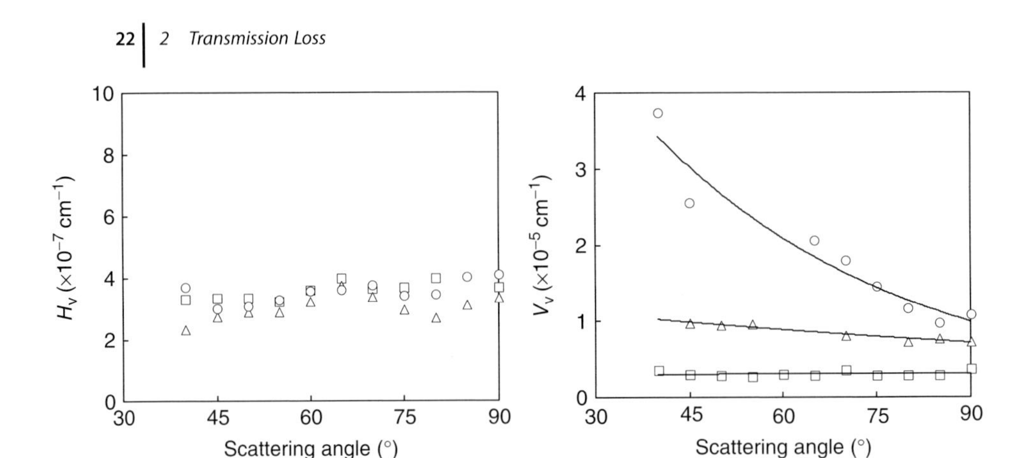

Figure 2.8 H_V and V_V scattering in PMMA glasses polymerized at 70 °C (○), 100 °C (△), and 130 °C (□) for 96 h.

and 2.20, the light scattering loss was estimated from the V_V^{iso} value to be only 9.5 dB/km. To explain the significant difference between the theoretical and experimental values, many theories have been proposed: stereo regularity according to the configuration of specific tacticities, the effect of high molecular weight, and the formation of cross-links. The origin of the excess light scattering was obscure. Subsequently, it has been clarified that the scattering losses in PMMA could be dramatically reduced by polymerizing or heating it above T_g [21, 22]. Figure 2.8 shows V_V and H_V for PMMA bulks polymerized at 70, 100, and 130 °C for 96 h. The wavelength of the incident light is 633 nm. As the polymerization temperature increases, the V_V intensity decreases, and no angular dependence is observed at 130 °C; in contrast, the H_V values are almost identical in the range of $3-4\times10^{-7}$ cm^{-1}. Their scattering losses, calculated using Debye's scattering method, are summarized in Table 2.2. Note that the α_1^{iso} value for PMMA polymerized at 130 °C is 9.7 dB/km; this is almost identical to the value predicted by Einstein's fluctuation theory. Furthermore, after heat treatment above T_g, the other two bulks polymerized at 70 and 100 °C also showed no angular dependence in V_V; the α_1^{iso} values were approximately 10 dB/km. These results indicate that the intrinsic factor causing the excess scattering can be eliminated by merely polymerizing or heating above T_g. In other words, the speculations above are invalid for

Table 2.2 Scattering parameters of PMMA glasses polymerized at 70, 100, and 130 °C for 96 h.

Polymerization temperature (°C)	D (Å)	$\langle\eta^2\rangle$ (×10^{-8})	α_1^{iso} (dB/km)	α_2^{iso} (dB/km)	α^{aniso} (dB/km)	α_s (dB/km)
70	676	1.05	16.8	40.8	4.4	62.0
100	466	0.53	17.7	10.9	4.0	32.6
130	—	0	9.7	0	4.7	14.4

explaining the phenomenon. Hence, the origin of the excess scattering is currently believed to be heterogeneity formed by volume shrinkage during polymerization.

2.2.4
Empirical Estimation of Scattering Loss for Amorphous Polymers

The fluctuation theory for structureless liquids expressed by Equation 2.18 indicates that the isotropic scattering loss $V_{\mathrm{V}}^{\mathrm{iso}}$ decreases with the isothermal compressibility β and refractive index n. Because the isotropic scattering is generally considerably greater than the anisotropic scattering for most amorphous polymers, the total scattering loss is roughly dependent on the isothermal compressibility and refractive index. These values can be obtained by the corresponding measurements; however, it is also possible to estimate them from the chemical structure [23]. The procedure is summarized in Figure 2.9.

The intrinsic molecular volume V_{int} of monomer units for an amorphous polymer can be calculated from the atomic radius and bond length of the constituent atoms on the basis of the method developed by Slonimskii *et al.* [24]. When an atom $\boldsymbol{B}$ (atomic radius R) is bound to atom $\boldsymbol{B}_i$ (atomic radius R_i) with bond length d_i, the atomic volume $\Delta V(\boldsymbol{B})$ of atom $\boldsymbol{B}$ is given by

$$\Delta V(\boldsymbol{B}) = \left(\frac{4}{3}\right)\pi R^3 - \sum_i \left(\frac{1}{3}\right)\pi h_i^2(3R - h_i),$$
$$h_i \equiv R - \frac{(R^2 + d_i^2 - R_i^2)}{2d_i}. \tag{2.32}$$

If the molecule consists of atoms $\boldsymbol{B}_1 - \boldsymbol{B}_j$, the intrinsic molecular volume V_{int} is given by

$$V_{\mathrm{int}} = N_{\mathrm{A}} \sum_j \Delta V(\boldsymbol{B}_j), \tag{2.33}$$

where N_{A} is Avogadro's number. The actual molecular volume of the monomer unit V, which contains the free volume, is expressed as

$$V = \frac{V_{\mathrm{int}}}{K}. \tag{2.34}$$

Here, K is the packing coefficient of the molecule; its value for most amorphous polymers is known to be 0.68 [23]. The actual molecular volume of the monomer unit V is related to the molecular weight between chain entanglements M_{c} by [25]

$$M_{\mathrm{c}} = 18.3V^{1.67}. \tag{2.35}$$

The number of chain atoms between physical entanglements, N_{c}, is expressed as

$$N_{\mathrm{c}} = \frac{M_{\mathrm{c}}}{M_0} Z. \tag{2.36}$$

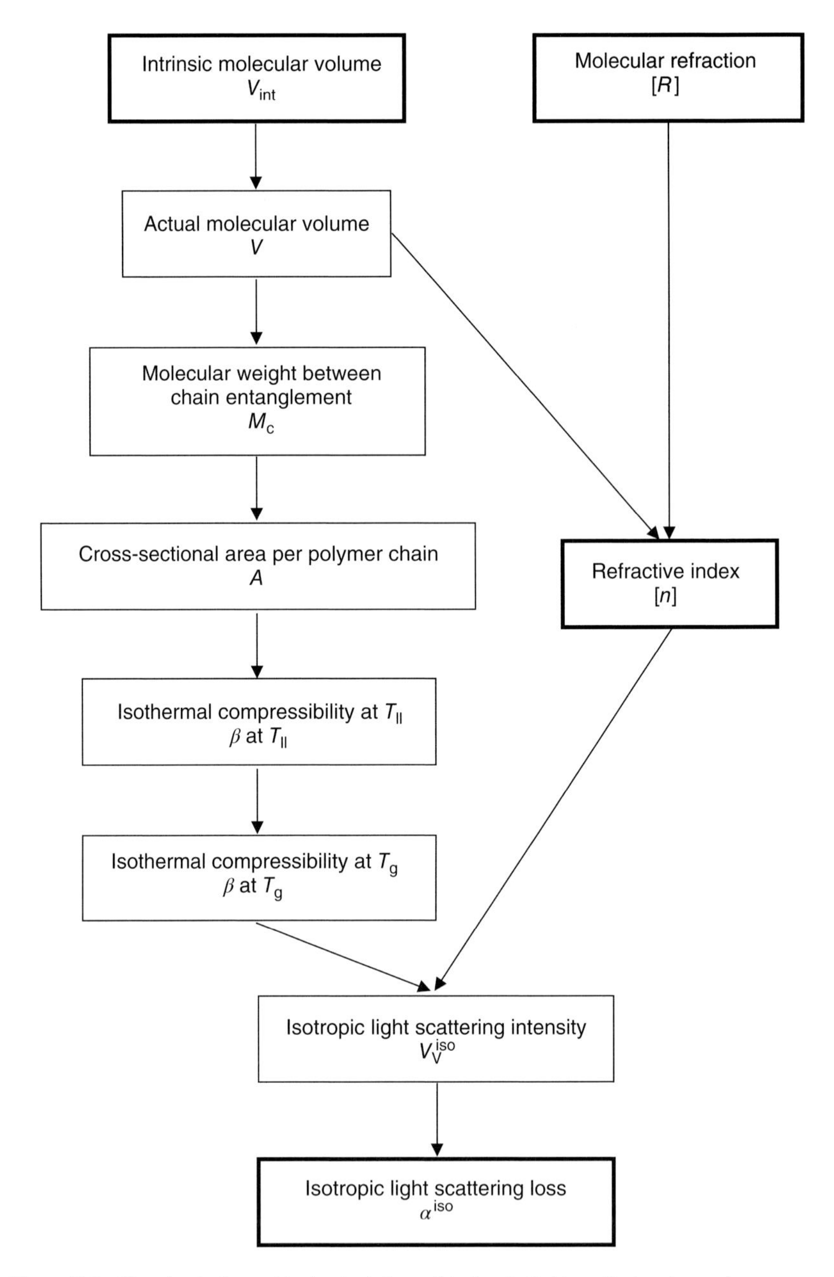

Figure 2.9 Flowchart of empirical calculation of isotropic light scattering losses in amorphous polymers.

Here, M_0 is the molecular weight of a monomer unit and Z is the number of chain atoms in the monomer unit. From the number of chain atoms between physical entanglements N_c, the cross-sectional area per polymer chain A is obtained as follows [26]:

$$\log N_c = k_1 + k_2(\log A - 2), \tag{2.37}$$

where k_1 and k_2 are constants ($k_1 = 2.929$, $k_2 = 0.614$).

Boyer and Miller reported the correlation between the isothermal compressibility β at the liquid–liquid transition temperature T_{ll} and the cross-sectional area per polymer chain A from the lattice parameters as [27]

$$\log(10^{11}\beta_{\text{at } T_{ll}}) = -0.21 + 0.55 \log A. \tag{2.38}$$

The liquid–liquid transition temperature appears to be a useful liquid-state reference temperature and is usually found by a variety of dynamic and thermodynamic methods. The concept of T_{ll} was vigorously studied by Boyer. He found that T_{ll} is near $1.2T_g$ for most polymers. Lobanov and Frenkel' [28] estimated T_{ll} for a variety of polymers from dielectric loss data and proposed an empirical relation between T_g and T_{ll}:

$$T_{ll} = T_g + 76\ (\text{K}). \tag{2.39}$$

For T_{ll} from 250 to 500 K, the numerical difference between the two empirical rules is small.

By combining the above relationships, the isothermal compressibility β of polymers at T_{ll} can be calculated from the intrinsic molecular volume V_{int}. However, the value of β at T_g is necessary to estimate the light scattering loss in the glassy state. β increases linearly with T from T_g to T_{ll}. It is interesting that the values of $(1/\beta_{\text{at } T_{ll}})(d\beta/dT)$ between T_g and T_{ll} for many amorphous polymers are almost the same (the average value is $4.8 \times 10^{-3}\ \text{K}^{-1}$) [23]. Considering the value $(1/\beta_{\text{at } T_{ll}})(d\beta/dT) = 4.8 \times 10^{-3}\ \text{K}^{-1}$ at $T_g < T < T_{ll}$ and the empirical relationship between T_{ll} and T_g ($T_{ll} = T_g + 76$ K), we can estimate β at T_g from β at T_{ll}.

The refractive index of a compound can be calculated from its molar refraction and molecular volume using the Lorentz–Lorenz equation

$$n = \sqrt{\left(2\frac{[R]}{V} + 1\right) / \left(1 - \frac{[R]}{V}\right)}, \tag{2.40}$$

where $[R]$ is the molar refraction and V is the actual molecular volume. $[R]$ is taken as the sum of the atomic refractions. Thus, we can estimate the refractive index of a polymer from the molar refraction and molecular volume.

Table 2.3 shows the calculated and observed values of the isothermal compressibility, refractive index, and isotropic light scattering of PMMA and PSt. The calculated values for both polymers show good agreement with the measured values.

Table 2.3 Isotropic light scattering in PMMA and PSt at 633 nm. Comparison of calculations from their molecular structures with measurements.

Polymer	β at Tg ($\times 10^{-11}$ cm^2/dyn)	n	V_V^{iso} ($\times 10^{-6}$ /cm)		α^{iso} (dB/km)	
	Calculated	Calculated	Calculated	Observed	Calculated	Observed
PMMA	3.6	1.494	2.7	2.7	9.8	9.7
PSt	4.4	1.583	5.6	5.9	20.4	21.5

2.3 Low-Loss POFs

2.3.1 PMMA- and PSt-Based POFs

The first POF, Crofon™, which was invented in the mid-1960s by DuPont, was a multimode fiber with a step-index (SI) profile in the core region. Compared to GOFs, the SI POF was also advantageous in terms of mass production; it was not only inexpensive to fabricate but also easy to mold and manufacture. The first commercial POF was Eska™, a PMMA-based SI POF introduced by Mitsubishi Rayon in 1975 [29]; subsequently, Asahi Chemical and Toray also entered the market. However, the first fibers were insufficiently transparent for use as an intra-building communication medium, and their application was severely limited to extremely short-range areas such as light guides, illuminations, audio data links, and sensors.

Through the analyses of each factor in fiber attenuation as described above, the limitations of POFs based on several materials and their theoretical grounding have been continually clarified. Further, it has been shown that the major factors causing high attenuation were not intrinsic but rather extrinsic, such as contaminants becoming mixed into the polymer during fiber fabrication. By preparing the fiber in an all-closed system from monomer distillation to fiber drawing, low-loss POFs with losses near the theoretical limits were fabricated. In 1981, Kaino reported a PSt-based SI POF with an attenuation of 114 dB/km at 670 nm [30], and in the following year, his group also succeeded in obtaining a PMMA-based SI POF with an attenuation of 55 dB/km at 568 nm [2].

2.3.2 PMMA-d$_8$-Based POF

Since the first SI POF was commercialized, most POFs have been manufactured using PMMA, a mass-produced, commercially available polymer that demonstrates high light transmittance and provides excellent corrosion resistance to both chemicals and weather. These properties, coupled with its low manufacturing costs and easy processing, have made PMMA a valuable substitute for glass in optical fibers. However, the performance can be further improved by modifying the base material. The high attenuation in conventional PMMA-based POFs is dominated by C–H overtone stretching and combinations of stretching and deformation. Hence, the most effective method of obtaining a lower loss POF is to replace hydrogen with heavier atoms such as fluorine, as shown in Figure 2.4. However, if vinyl monomers such as MMA are perfluorinated, the polymerization rate dramatically decreases.

In 1977, DuPont developed an SI POF based on perdeuterated PMMA (PMMA-d_8) [31]. The attenuation spectrum showed low-loss windows at longer wavelengths such as 690 and 790 nm compared to those of PMMA. This is because the overtones of molecular vibrational absorption shifted to longer wavelengths. However, the minimum attenuation of the fiber was still 180 dB/km, which was even higher than that of PMMA at the time. One of the advantages of deuteration is that deuterium efficiently reduces the absorption loss at visible to near-infrared wavelengths but does not affect most other optical and physical properties because deuterium is an isotope of hydrogen. In other words, the light scattering intensity of PMMA-d_8 should be similar to that of PMMA, as discussed in Section 2.2.3. Kaino believed that the high attenuation in the PMMA-d_8-based SI POF was caused by extrinsic factors such as scattering losses due to dust and microvoids and absorption losses due to inorganic contaminants such as transition-metal ions. Thus, he prepared the fiber using the all-closed extrusion system that was used to fabricate the low-loss PMMA-based POF and succeeded in decreasing the attenuation to 20 dB/km at 650–680 nm [32]. In addition, the polymer was extensively studied as a base material for GI (graded-index) POFs, and an attenuation of 80 dB/km at 650 nm was achieved in 2005 [33].

2.3.3 *CYTOP*®-Based POF

CYTOP®, developed and trademarked by AGC, is an amorphous perfluorinated polymer. The first CYTOP-based GI POF, Lucina®, was commercialized in 2001. Figure 2.10 compares the attenuation spectra of PMMA-, PMMA-d_8-, and CYTOP-based GI POFs. CYTOP molecules consist solely of C–C, C–F, and C–O bonds. The wavelengths of the fundamental stretching vibrations of these atomic bonds are relatively long; therefore, the vibrational absorption loss in CYTOP at the light source wavelengths is negligibly small. In addition, it exhibits fairly low light scattering because of its low refractive index ($n_D = 1.34$). The attenuation in a CYTOP-based GI POF is approximately 10 dB/km at 1.0 μm. The theoretical attenuation spectrum of a CYTOP-based GI POF was calculated using the Morse potential energy theory and thermally induced fluctuation theory to consider the inherent absorption and scattering losses, respectively [34]. Given that the theoretical limit of attenuation is 0.7 dB/km at this wavelength, it is expected that the attenuation can be lowered further by preventing contamination during fabrication.

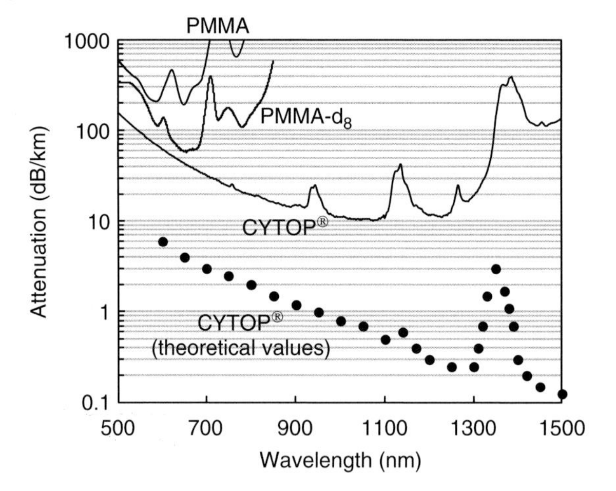

Figure 2.10 Attenuation spectra of PMMA-, PMMA-d_8-, and CYTOP-based GI POFs.

References

1. Urbach, F. (1953) The long-wavelength edge of photographic sensitivity and of the electronic absorption of solids. *Phys. Rev.*, **92** (5), 1324.
2. Kaino, T., Fujiki, M., and Jinguji, K. (1984) Preparation of plastic optical fibers. *Rev. Electr. Commun. Lab.*, **32** (3), 478–488.
3. Yamashita, T. and Kamada, K. (1993) Intrinsic transmission loss of polycarbonate core optical fiber. *Jpn. J. Appl. Phys.*, **32** (6A), 2681–2686.
4. Atkins, P. and Friedman, R. (2005) *Molecular Quantum Mechanics*, Oxford University Press, New York.
5. Gough, K.M. and Henry, B.R. (1984) Overtone spectral investigation of substituent-induced bond-length changes in gas-phase fluorinated benzenes and their correlation with ab initio STO-3G and 4-21G calculations. *J. Am. Chem. Soc.*, **106** (10), 2781–2787.
6. Henry, B.R. (1987) The local mode model and overtone spectra: a probe of molecular structure and conformation. *Acc. Chem. Res.*, **20** (12), 429–435.
7. Snavely, D.L., Blackburn, F.R., Ranasinghe, Y., Walters, V.A., and Gonzalez del Riego, M. (1992) Vibrational overtone spectroscopy of pyrrole and pyrrolidine. *J. Phys. Chem.*, **96** (9), 3599–3605.
8. Timme, B. and Mecke, R. (1936) Quantitative absorptionsmessungen an den CH-Obserschwingungen einfacher Kohlenwasserstoffe. *Z. Phys.*, **98** (5), 363–381.
9. Groh, W. (1988) Overtone absorption in macromolecules for polymer optical fibers. *Makromol. Chem.*, **189** (12), 2861–2874.
10. Ballato, J., Foulger, S.H., and Smith, D.W. Jr., (2004) Optical properties of perfluorocyclobutyl polymers. II. Theoretical and experimental attenuation. *J. Opt. Soc. Am. B*, **21** (5), 958–967.
11. Koike, K. and Koike, Y. (2009) Design of low-loss graded-index plastic optical fiber based on partially fluorinated methacrylate polymer. *J. Lightwave Technol.*, **27** (1), 41–46.
12. Yen, C.-T. and Chen, W.-C. (2003) Effects of molecular structures on the near-infrared optical properties of polyimide derivatives and their corresponding optical waveguides. *Macromolecules*, **36** (9), 3315–3319.
13. Ghim, J., Lee, D.-S., Shin, B.G., Vak, D., Yi, D.K., Kim, M.-J., Shim, H.-S., Kim, J.-J., and Kim, D.-Y. (2004) Optical properties of perfluorocyclobutane aryl ether

polymers for polymer photonic devices. *Macromolecules*, **37** (15), 5724–5731.
14. Takahashi, A., Inoue, A., Sassa, T., and Koike, Y. (2013) Fluorination effects on attenuation spectra of plastic optical fiber core materials. *Opt. Mater. Express*, **3** (5), 658–663.
15. Olivero, J.J. and Longbothum, R.L. (1977) Empirical fits to the Voigt line width: a brief review. *J. Quant. Spectrosc. Radiat. Transfer*, **17** (2), 233–236.
16. Liu, Y., Lin, J., Huang, G., Guo, Y., and Duan, C. (2001) Simple empirical analytical approximation to the Voigt profile. *J. Opt. Soc. Am. B*, **18** (5), 666–672.
17. Einstein, A. (1910) Theorie der Opaleszenz von homogenen Flüssigkeiten und Flüssigkeitsgemischen in der Nähe des kritischen Zustandes. *Ann. Phys.*, **338** (16), 1275–1298.
18. Fischer, E.W. and Dettenmaier, M. (1976) Untersuchungen zu den orientierungs- und dichtefluktuationen in amorphen polymeren mit hilfe der lichtstreuung. *Makromol. Chem.*, **177** (4), 1185–1197.
19. Debye, P. and Bueche, A.M. (1949) Scattering by an inhomogeneous solid. *J. Appl. Phys.*, **20** (6), 518–525.
20. Hellwege, K.-H., Knappe, W., and Lehmann, P. (1962) Die isotherme Kompressibilität einiger amorpher und teilkristalliner Hochpolymerer im Temperaturbereich von 20–250 °C und bei Drucken bis zu 2000 kp/cm^2. *Kolloid Z. Z. Polym.*, **183** (2), 110–120.
21. Koike, Y., Tanio, N., and Ohtsuka, Y. (1989) Light scattering and heterogeneities in low-loss poly(methyl methacrylate) glasses. *Macromolecules*, **22** (3), 1367–1373.
22. Koike, Y., Matsuoka, S., and Bair, H.E. (1992) Origin of excess light scattering in poly(methyl methacrylate) glasses. *Macromolecules*, **25** (18), 4807–4815.
23. Tanio, N. and Koike, Y. (1997) Estimate of light scattering loss of amorphous polymer glass from its molecular structure. *Jpn. J. Appl. Phys.*, **36** (1–2), 743–748.
24. Slonimskii, G.L., Askadskii, A.A., and Kitaigorodskii, A.I. (1970) On the macromolecular packing in polymers. *Vysokomol. Soed.*, **A12** (3), 494–512.
25. Hoffmann, M. (1972) Fließverhalten und molekulare struktur von polymeren und ihren lösungen eine systematik quantitativer beziehungen. *Makromol. Chem.*, **153** (1), 99–124.
26. Boyer, R.F. and Miller, R.L. (1978) Chain entanglements and chain areas II: a molecular basis for chain entanglements. *Rubber Chem. Technol.*, **51** (4), 718–730.
27. Boyer, R.F. and Miller, R.L. (1984) Correlation of liquid-state compressibility and bulk modulus with cross-sectional area per polymer chain. *Macromolecules*, **17** (3), 365–369.
28. Lobanov, A.M. and Frenkel', S.Y. (1980) The nature of the so-called "liquid-liquid" transition in polymer melts. *Polym. Sci. USSR*, **22** (5), 1150–1163.
29. Mitsubishi Rayon Co. Ltd. (1975) JP Patent 1975-83046, filed Nov. 22, 1973 and issued Jul. 4, 1975.
30. Kaino, T., Fujiki, M., and Nara, S. (1981) Low-loss polystyrene core-optical fibers. *J. Appl. Phys.*, **52** (12), 7061–7064.
31. Schleinitz, H.M. (1977) Ductile plastic optical fibers with improved visible and near infrared transmission. Proceeding of 26th International Wire and Cable Symposium, CECOM, Ft. Monmouth, NJ, November 15–17, pp. 352–355.
32. Kaino, T., Jinguji, K., and Nara, S. (1983) Low loss poly(methyl methacrylate-d8) core optical fibers. *Appl. Phys. Lett.*, **42** (7), 567–569.
33. Kondo, A., Ishigure, T., and Koike, Y. (2005) Fabrication process and optical properties of perdeuterated graded-index polymer optical fiber. *J. Lightwave Technol.*, **23** (8), 2443–2448.
34. Tanio, N. and Koike, Y. (2000) What is the most transparent polymer? *Polym. J.*, **32** (1), 43–50.

3
Transmission Capacity

The bandwidth of a fiber determines the maximum transmission data rate or maximum transmission distance. Most common POF (plastic optical fiber) transmission systems adopt on – off keying by direct modulation of the optical source (laser or light-emitting diode). If an input pulse waveform can be detected without distortion at the other end of the fiber, the maximum link length is limited by the fiber attenuation. However, in addition to the optical power attenuation, the output pulse is generally broader in time than the input pulse. This pulse broadening limits the transmission capacity, namely, the bandwidth of the fiber. The bandwidth is determined by the impulse response as follows [1]: Optical fibers are usually considered quasi-linear systems; thus, the output pulse is described by

$$p_{out}(t) = h(t) * p_{in}(t). \tag{3.1}$$

The output pulse $p_{out}(t)$ from the fiber can be calculated in the time domain through the convolution (denoted by *) of the input pulse $p_{in}(t)$ and the impulse response function $h(t)$ of the fiber. Fourier transformation of Equation 3.1 provides a simple expression as the product in the frequency domain:

$$P_{out}(f) = H(f)P_{in}(f), \tag{3.2}$$

where $H(f)$ is the power transfer function of the fiber at the baseband frequency f. The power transfer function defines the bandwidth of the optical fiber as the lowest frequency at which $H(f)$ is reduced to half the DC value. The power transfer function is easily calculated from the Fourier transform of the experimentally measured input and output pulses in the time domain, or from the measured output power from the fiber in the frequency domain from DC to the bandwidth frequency. A higher bandwidth yields less pulse broadening and enables higher speed data transmission. The bandwidth limitation also largely determines the maximum link length for a given data rate in some multimode fiber (MMF) systems. Pulse broadening, which is theoretically proportional to the fiber length in the absence of mode coupling, is caused mainly by two dispersion mechanisms in POFs: intermodal and intramodal dispersion. The root mean squared (rms) width of the impulse response σ is calculated by $\sigma = \sqrt{\sigma_{inter}^2 + \sigma_{intra}^2}$, where σ_{inter} and σ_{intra} are the rms widths of the pulse broadening induced by intermodal and intramodal dispersion, respectively. Another type of dispersion, called *polarization mode dispersion*, arises from the anisotropies of the structure and material,

Fundamentals of Plastic Optical Fibers, First Edition. Yasuhiro Koike.

which results in slightly different propagation constants for the two orthogonal polarization modes [2]. The effect of polarization mode dispersion can usually be ignored in MMFs. Therefore, this chapter explains the intermodal and intramodal dispersion and the bandwidths of POFs.

3.1 Bandwidth

3.1.1 Intermodal Dispersion

When an optical pulse is input into an MMF, the optical power of the pulse is generally distributed into the huge number of modes of the fiber. Different modes travel at different propagation speeds along the fiber; thus, different modes launched at the same time reach the output end of the fiber at different times. Therefore, the input pulse is broadened in time as it travels along the MMF. This pulse broadening effect, well known as *modal dispersion*, is significantly observed in step-index (SI) MMFs. Different rays travel along paths with different lengths, where each distinct ray can be thought of as a mode in a simple interpretation. The light travels at the same velocity along its optical path because of the constant refractive index throughout the core region in SI MMFs. Consequently, the same velocity but different path lengths result in different propagation speeds along the fiber, which causes a pulse spread in time. The pulse broadening caused by modal dispersion seriously limits the transmission speed in MMFs because overlapping of the broadened pulses induces intersymbol interference and disrupts correct signal detection, thereby increasing the bit error rate [3].

Modal dispersion is generally a dominant factor in pulse broadening in MMFs, and in particular in SI MMFs. As shown in [4] for an ideal SI MMF, the difference in the propagation delay between the fastest and slowest modes is given by

$$\Delta t_{\mathrm{mod}} = \frac{L}{2cn_2} NA^2, \tag{3.3}$$

where L is the fiber length, n_1 and n_2 are the core and cladding refraction indices, respectively, and NA = numerical aperture = $\sqrt{n_1^2 - n_2^2}$. For instance, for a typical NA of 0.5, after 50 m of SI POF, one would obtain a Δt_{mod} of 14.8 ns, and thus an estimated electrical-to-electrical available bandwidth of $B \cong 0.44/\Delta t_{\mathrm{mod}} \approx 30\,\mathrm{MHz}$ under worst case conditions. The actual bandwidth depends on several other parameters, related mainly to the NA of the source, the level of mode mixing inside the POF, and the fact that higher order modes (which travel more slowly) usually see higher attenuation and thus contribute less to the overall transfer function (see again [4]) than lower order modes. As a result, the actual available bandwidths for SI POFs are higher than those given by the theoretical treatment above, but are still limited to less than approximately 100 MHz over 50 m.

However, the modal dispersion can be dramatically reduced by forming a near-parabolic refractive index profile in the core region of the MMF, which allows a much higher bandwidth (i.e., higher-speed data transmission) [5]. A typical graded-index (GI) MMF has a cylindrically symmetric refractive index profile that gradually decreases from the core axis to the core–cladding interface. A ray confined near the core axis, corresponding to a lower order mode, travels a shorter geometrical length at a slower light velocity along the path because of the higher refractive index. A sinusoidal ray passing through near the core–cladding boundary, considered as a higher order mode, travels a longer geometrical length at a faster velocity along the path, particularly in the lower refractive index region far from the core axis. As a result, the output times from the fiber end of light traveling the shorter geometrical length at the slower velocity and the longer geometrical length at the faster velocity can be made almost the same by using the optimum refractive index profile. Thus, the refractive index distribution strongly affects the bandwidth of MMFs. The optimization of the refractive index profile in GI MMFs is an important issue in reducing the modal dispersion. A power-law index profile approximation is a well-known method of analyzing the optimum profile of GI MMFs [6]. In this approximation, the refractive index distribution of a GI MMF is described by

$$n(R) = \begin{cases} n_1 \left[1 - 2\Delta\left(\frac{R}{a}\right)^g\right]^{\frac{1}{2}} & \text{for} \quad 0 \le R \le a \\ n_2 & \text{for} \quad R > a \end{cases}, \tag{3.4}$$

where $n(R)$ is the refractive index as a function of the radial distance R from the core center, n_1 and n_2 are the refractive indices of the core center and the cladding, respectively, and a is the core radius. The profile exponent g determines the shape of the refractive index profile, and Δ is the relative index difference, given by

$$\Delta = \frac{n_1^2 - n_2^2}{2n_1^2}. \tag{3.5}$$

Equation 3.4 includes the SI profile when $g = \infty$.

The optimum profile exponent g_{opt}, which minimizes the modal dispersion and the difference in the delay of all the modes and maximizes the bandwidth, is expressed as follows on the basis of an analysis of scalar wave equation:

$$g_{opt} = 2 + \eta - \Delta\frac{(4+\eta)(3+\eta)}{5+2\eta}, \tag{3.6}$$

$$\eta = -\frac{2n_1}{N_1}\frac{\lambda}{\Delta}\frac{d\Delta}{d\lambda}, \quad N_1 = n_1 - \lambda\frac{dn_1}{d\lambda} \tag{3.7}$$

If the refractive index of the material is wavelength-independent, Equation 3.6 becomes the simple expression

$$g_{opt} = 2 - \frac{12}{5}\Delta. \tag{3.8}$$

A high bandwidth can typically be achieved when the profile exponent g is approximately equal to 2.0. However, the refractive indices of materials generally depend on the wavelength, which induces profile dispersion.

The profile dispersion p caused by the wavelength dependence of the refractive index [7] is given by

$$p = \frac{\lambda}{\Delta} \frac{d\Delta}{d\lambda}. \tag{3.9}$$

The g_{opt} value depends on the relative index difference Δ, which is a function of the refractive indices of the core and the cladding. These refractive indices are determined by the wavelength and the dopant characteristics. If the refractive indices of the polymer matrix and dopant have identical wavelength dependences, g_{opt} obeys Equation 3.8. However, the wavelength dependence of the refractive index of the dopant generally differs from that of the polymer matrix; hence, the shape of the refractive index profile depends on the wavelength. Therefore, even if the optimum refractive index profile is provided at a certain wavelength, this profile differs from the optimum profile at another wavelength. Profile dispersion also depends on the wavelength of the light signal. The effect of profile dispersion can be compensated in the refractive index profile by taking it into account, which is easily illustrated by Equations 3.6, 3.7, and 3.9. Thus, the g_{opt} value is shifted by profile dispersion. Consequently, the intermodal dispersion can be minimized by using the optimum refractive index profile considering profile dispersion. Then, intramodal dispersion becomes important for achieving high bandwidth. Additionally, single-mode fibers (SMFs) exhibit even higher bandwidths than GI MMFs because modal dispersion does not exist in principle; thus, intramodal dispersion seriously limits the bandwidth of SMFs [8].

3.1.2
Intramodal Dispersion

Intramodal dispersion or chromatic dispersion is the pulse widening caused by the finite spectral width of the light source. Intramodal dispersion comprises material and waveguide dispersion. Material dispersion is induced by the wavelength dependence of the refractive index of the core material [9]. The group velocity of a given mode depends on the wavelength; thus, the output pulse is broadened in time even when optical signals with different wavelengths follow the same path. This effect is generally much smaller than modal dispersion in MMFs but is no longer negligible when the modal dispersion is sufficiently suppressed.

Waveguide dispersion arises from the wavelength dependence of the optical power distribution of a mode between the core and the cladding. Light at shorter wavelengths is more completely confined to the core region, and light at longer wavelengths is distributed more in the cladding. Because more longer wavelength light is in the cladding, it travels at a higher propagation speed, as the refractive index of the cladding is lower than that of the core.

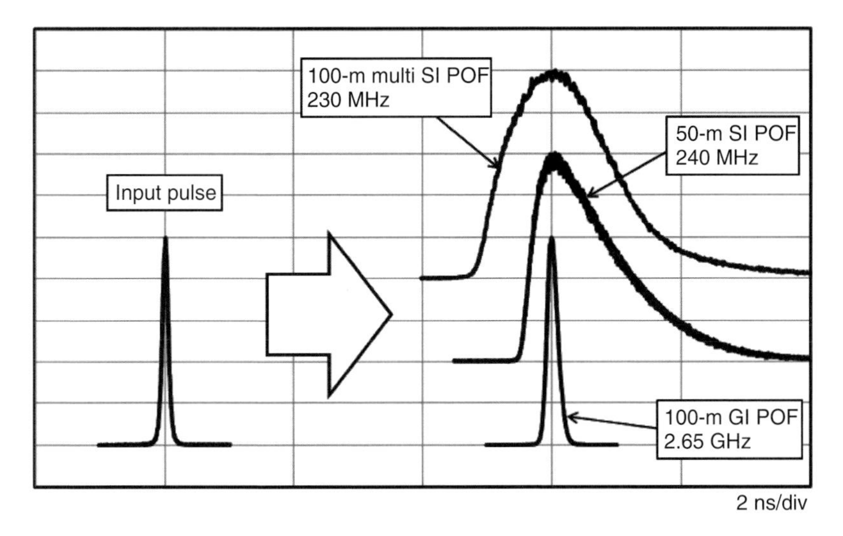

Figure 3.1 Measured output pulse waveforms through 50-m SI POF, 100-m multi-SI POF, and 100-m GI POF under overfilled launch condition.

3.1.3
High-Bandwidth POF

A large number of modes, typically more than tens of thousands, can propagate in POFs because of the large cores. Therefore, reducing modal dispersion has been a key issue in such multimode POFs. The measured output pulse waveforms from poly(methyl methacrylate)- (PMMA-) based SI, multi-SI, and GI POFs under overfilled launch conditions are shown in Figure 3.1 in comparison with the input pulse waveform. The bandwidth of the SI POF was seriously limited. The impulse responses and bandwidths of the SI and multi-SI POFs were almost the same, even though the multi-SI POF was twice as long as the SI POF. The multi-SI POF can realize communication at a data rate twice as high as that of the SI POF; however, the bandwidth was still limited to hundreds of megahertz for 100 m. The GI POF was proposed to reduce the bandwidth limitation caused by the large modal dispersion of the SI POF [10], and the refractive index profile can be precisely controlled and optimized. The bandwidth of the GI POF was dramatically enhanced to several gigahertz over 100 m by such control of the index profile. This is due to the decrease in modal dispersion caused mainly by the graded refractive index profile. Thus, the GI POF is a promising candidate for a transmission medium for high-speed and short-reach networks.

Figure 3.2 shows the effects of individual types of dispersion on the bandwidth characteristics of PMMA-based GI POFs. Here, the bandwidths were estimated by the Equations 3.10–3.12 [11, 12]. Gaussian pulses were assumed, and the rms spectral width was set to 1 nm.

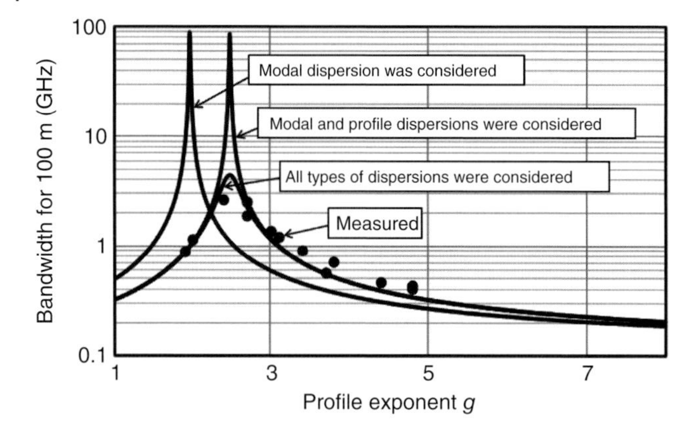

Figure 3.2 Calculated bandwidth of 100-m PMMA-based GI POF as a function of the profile exponent g at a wavelength of 650 nm considering individual types of dispersion compared with the measured bandwidths under overfilled launch conditions. An rms spectral width of 1 nm was assumed.

$$\sigma_{\text{inter}} = \frac{LN_1\Delta}{2c}\frac{g}{g+1}\left(\frac{g+2}{3g+2}\right)^{\frac{1}{2}}\left[C_1^2 + \frac{4C_1C_2\Delta\,(g+1)}{2g+1} + \frac{4\Delta^2C_2^2(2g+2)^2}{(5g+2)(3g+2)}\right]^{\frac{1}{2}}, \tag{3.10}$$

$$\sigma_{\text{intra}} = \frac{L\sigma_\lambda}{c\lambda}\left[\begin{array}{c}\left(-\lambda^2 n_1''\right)^2 - 2\lambda^2 n_1''(N_1\Delta)\left(\frac{g-2-\eta}{g+2}\right)\left(\frac{2g}{2g+2}\right)\\ +(N_1\Delta)^2\left(\frac{g-2-\eta}{g+2}\right)^2\frac{4g^2}{(g+2)(3g+2)}\end{array}\right], \tag{3.11}$$

$$C_1 = \frac{g-2-\eta}{g+2}, \quad C_2 = \frac{3g-2-2\eta}{2\,(g+2)}. \tag{3.12}$$

The maximum bandwidth, ~100 GHz for 100 m, was theoretically obtained when g was almost 2.0 if only the modal dispersion was considered. However, large discrepancies were observed between the calculated and measured bandwidths. In contrast, the relation between the bandwidth and the profile exponent of PMMA-based GI POFs was accurately estimated by the Wentzel–Kramers–Brillouin method when all the dispersion factors were taken into account.

On the other hand, the g_{opt} value, which represents the highest bandwidth, shifted to around 2.4 because of profile dispersion, as discussed above. The dopant used in all the GI POFs shown in Figure 3.2 was diphenyl sulfide, and all the GI POFs had NAs of ~0.20. This result indicates that the bandwidth characteristics are strongly influenced by the profile dispersion even if the refractive index profile deviates greatly from the ideal one.

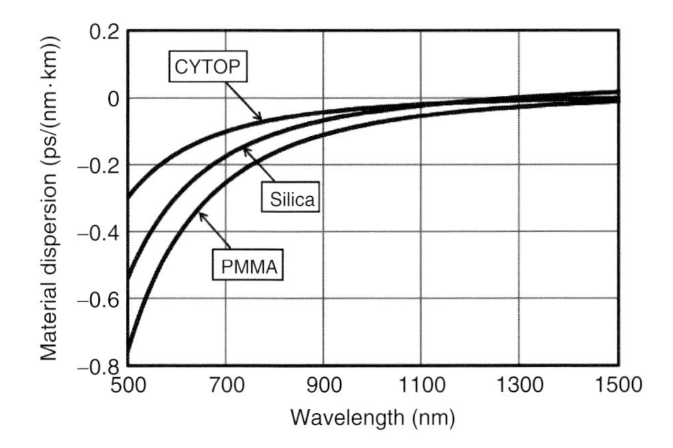

Figure 3.3 Material dispersion of PMMA, CYTOP, and silica.

The calculated maximum bandwidth at g_{opt} was dramatically reduced owing to the large material dispersion of PMMA when all types of dispersion were considered. A significant decrease in bandwidth was observed mainly for profile exponents near the g_{opt} value. When the profile exponent deviates greatly from g_{opt}, material dispersion has little effect on the bandwidth because of the dominant large modal dispersion.

The bandwidth of the PMMA-based GI POFs can be optimized by accurate control of the refractive index profile [13]. Further bandwidth improvement has been investigated by using perfluorinated (PF) polymers [14]. PF polymers have the valuable characteristics of low material dispersion and low transmission losses (see Chapter 2). Figure 3.3 shows the material dispersion in PMMA (a PF polymer commercially available from Asahi Glass Co.), and silica calculated from the wavelength dependence of their refractive indices. The material dispersion of CYTOP® is much smaller than that of PMMA and even that of silica, particularly in the visible to near-infrared region, which means that CYTOP-based GI POFs with an optimum refractive index profile can realize higher bandwidths than silica-based GI MMFs. The material dispersion curve of CYTOP was insensitive to the wavelength, unlike those of silica and PMMA. Hence, the bandwidth of CYTOP-based GI POFs is also expected to be insensitive to the wavelength. Thus, CYTOP-based GI POFs are predicted to have a higher bandwidth than silica-based GI MMFs. The PF GI POFs, indeed, demonstrated data transmission rates of 40 Gb/s and even higher over 100 m [15–17].

The wavelength dependence of the bandwidths of 100-m PMMA- and CYTOP-based GI POFs and a silica-based GI MMF was theoretically estimated from the wavelength dependence of their refractive indices, as shown in Figure 3.4. The refractive index profiles of all the fibers were optimized at a wavelength of 850 nm, and the rms spectral width of the light source was assumed to be 1.0 nm. The bandwidth at 850 nm of the PMMA-based GI POF, almost 10 GHz, was higher than that at 650 nm because of the lower material dispersion, although a light signal at 850 nm cannot travel a long distance in PMMA because of the

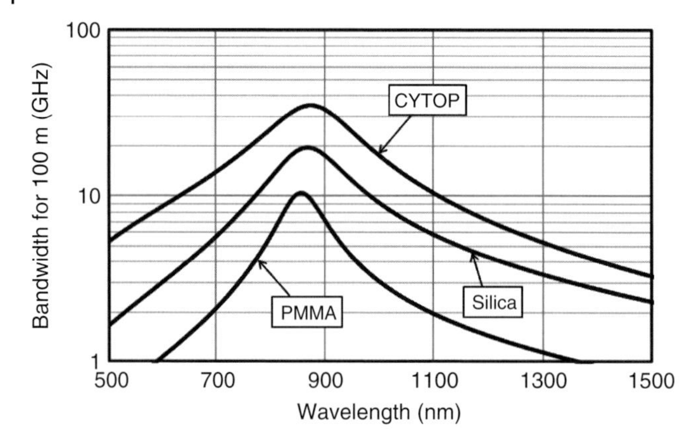

Figure 3.4 Calculated bandwidths of PMMA- and CYTOP-based GI POFs and silica-based GI MMF as a function of wavelength. Refractive index profiles were optimized at a wavelength of 850 nm, and the rms spectral width of light source was assumed to be 1 nm.

large transmission loss. The wavelength dependence of the optimum profile of the CYTOP-based GI POF was small, as mentioned above, because CYTOP has low material and profile dispersions. Therefore, the CYTOP-based GI POF can maintain its high-bandwidth characteristics over a wide wavelength range compared to silica-based GI MMFs. Consequently, CYTOP-based GI POFs can use various light sources with a larger variety of wavelengths; hence, more channels are available in wavelength division multiplexing systems. This indicates that CYTOP-based GI POFs can realize higher data rate transmission systems.

3.2
Wave Propagation in POFs

The transmission bandwidths of GI POFs have been extended with the development of low-dispersive materials and GI profile control techniques. Consequently, a 40-Gbps serial transmission over 100 m of GI POFs has been demonstrated. It has also been reported that the transmission characteristics of GI POFs are significantly affected by much stronger mode coupling than that in glass MMFs [18–20], although the underlying physics is not well understood. This suggests that mode coupling in GI POFs can originate in some phenomenon other than microbending, the predominant origin in glass MMFs, whose mode coupling effects can typically be observed for transmission over several kilometers [6]. Therefore, mode coupling in GI POFs cannot be analyzed using the conventional diffusion theory of coupled power equations [21].

Optical fiber materials exhibit density fluctuations because of their amorphous nature. These fluctuations in glasses result in Rayleigh scattering because their characteristic scales are much smaller than the wavelengths of guided light used in glass MMFs. Rayleigh scattering induces random mode coupling, but the effect is too weak to influence the transmission bandwidth of glass MMFs with lengths of less than several hundred meters. However, polymers such as PMMA have longer scale fluctuations of microscopic heterogeneous structures [22, 23]. Therefore, microscopic heterogeneities in POF cores can result in mode coupling because the POF cores exhibit greater forward scattering than glasses. However, the intrinsic material properties have been overlooked in the transmission analyses of GI POFs. Here, we clarify the effect of the intrinsic microscopic properties on the macroscopically observed transmission characteristics of GI POFs. Moreover, we show that intrinsic mode coupling results in significant noise reduction in MMF optical links, which allows low-cost radio-over-fiber (RoF) systems in home and building networks.

3.2.1 Microscopic Heterogeneities

As mentioned in Chapter 2, attenuation in GI POFs has been sufficiently reduced for short-reach applications, according to our study of light scattering in GI POF core bulk polymers. However, the remaining scattering losses in GI POFs are much higher than those in glass MMFs. Actually, we can visually observe streaking of the light trajectory in a PMMA-based GI preform because of the intrinsic forward scattering, as shown in Figure 3.5. This indicates that GI POF cores can have microscopic heterogeneities in their higher order structures, which can result in stronger mode coupling because greater forward scattering occurs than in glass MMFs.

The microscopic heterogeneities in actual GI POF cores can be influenced by the fabrication processes, and their correlation characteristics differ significantly from those of the polymer bulk. In GI POFs, moreover, guided light waves can interact with microscopic heterogeneities through fluctuations in the cross-sectional components of the dielectric constants along the fiber axis because the guided wave fronts are almost parallel to the cross section in GI POFs under the weakly guiding approximation. This makes the experimental evaluation of the microscopic

Figure 3.5 Visually observable light trajectory due to forward scattering by microscopic heterogeneous structures in PMMA-based GI preform.

heterogeneities in GI POF cores difficult. Therefore, light scattering by microscopic heterogeneities in actual GI POF cores has not been well understood.

3.2.2
Debye's Scattering Theory

Light scattering by microscopic heterogeneities can be understood in terms of the Rayleigh–Debye scattering theory [24], which Debye *et al.* [25, 26] applied to problems regarding the characterization of inhomogeneities in solids by light or X-ray scattering measurements. According to their methods, the light scattering intensity pattern can be related to the spatial correlation characteristics of the dielectric constant fluctuations, which reflect the microscopic material properties. Here, the light scattering theory is briefly summarized to understand light scattering by microscopic heterogeneities in polymers.

Let us consider monochromatic light scattered from an inhomogeneous solid with an average dielectric constant ε_{av} on which are superimposed local fluctuations $\delta\varepsilon$. For a weakly scattering medium, where the fluctuation is small compared to the average, the single-scattering approximation can be applied to a scattered light field because the incident light is perturbed little by the fluctuation [24, 27]. In the Rayleigh–Debye approximation, the scattered field amplitude in the far zone can be expressed as a superposition of contributions from volume elements:

$$F(\boldsymbol{s}) = \int \delta\varepsilon(r) e^{-ik\boldsymbol{s}\cdot r} dr, \tag{3.13}$$

with

$$\boldsymbol{s} = \boldsymbol{S} - \boldsymbol{S}_0. \tag{3.14}$$

Here, $\boldsymbol{S}$ and $\boldsymbol{S}_0$ are the unit wave vectors of scattered and incident light, respectively, and k is the light wavenumber $2\pi/\lambda$, where λ is the wavelength. The amplitude of $F(\boldsymbol{s})$ without the proportionality factor is analogous to the form factor of an atom in X-ray scattering. A photodetector measures not the scattered light amplitude but the scattered light intensity, which is proportional to a time average of the $|F(\boldsymbol{s})|^2$ value. For amorphous materials, the dielectric constant fluctuations can be considered to be density fluctuations that are frozen in at the glass transition temperature. Assuming that the dynamic fluctuation is a stationary ergodic process, the scattering light intensity can be expressed as

$$I(\boldsymbol{s}) = \langle |F(\boldsymbol{s})|^2 \rangle = V \int \gamma(r) e^{ik\boldsymbol{s}\cdot r} dr, \tag{3.15}$$

where the angle brackets represent the ensemble average. V is the irradiated volume and $\gamma(r)$ is the autocorrelation function, which is defined as the ensemble-averaged product of the fluctuation $\delta\varepsilon_1$ and $\delta\varepsilon_2$ at two points with a separation distance of r. In the derivation of Equation 3.15, we assumed an isotropic fluctuation whose autocorrelation function depends only on the distance r. Therefore, the following expression is also obtained by performing the integral in Equation 3.15 over all directions:

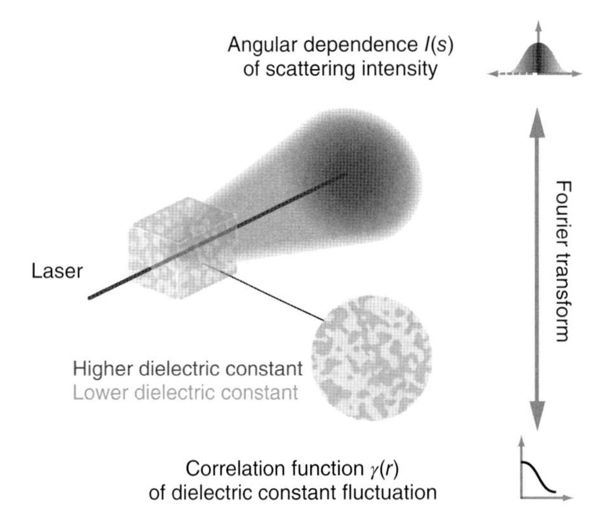

Figure 3.6 Light scattering by microscopic heterogeneities.

$$I(\mathbf{s}) = V \int_0^{\infty} 4\pi r^2 \gamma(r) \frac{\sin(ksr)}{ksr} dr, \tag{3.16}$$

where $s = |\mathbf{s}|$. Equation 3.15 shows that the scattering intensity pattern of the polymers is related to the autocorrelation function of microscopic heterogeneities by the Fourier transform, as shown in Figure 3.6. Assuming Gaussian-correlated heterogeneities, the autocorrelation function is given by

$$\gamma(r) = \langle \delta\varepsilon^2 \rangle \exp\left(-\frac{r^2}{D^2}\right), \tag{3.17}$$

where the correlation length D and mean square fluctuation $\langle \delta\varepsilon^2 \rangle$ correspond to measures of the fluctuation size and amplitude, respectively. The fluctuation size or correlation length determines the angular scattering profile, and the fluctuation amplitude controls the scattering efficiency. From Equations 3.15 and 3.17, we obtain the intensity of light scattered from the microscopic heterogeneities, as

$$I(\mathbf{s}) = V \langle \delta\varepsilon^2 \rangle \frac{4\pi^{\frac{3}{2}} D^3 e^{-\frac{k^2 s^2 D^2}{4}}}{4 + k^2 s^2 D^2}. \tag{3.18}$$

A longer correlation length results in more forward scattering, as shown in Equation 3.18. Microscopic heterogeneities have a larger scale correlation of the dielectric constant fluctuation than those in optical glasses. Therefore, the optical transmission characteristics of POFs differ inherently from those of glass MMFs through random mode coupling due to greater forward scattering by the microscopic heterogeneities.

3.2.3 Developed Coupled Power Theory

Microscopic heterogeneities in POF core materials result in random power transitions between the guided modes through forward scattering, as illustrated in Figure 3.7. Mode coupling can affect the mode power distribution and thus the transmission bandwidth [28]. Power transitions from guided modes to radiation modes are also induced; this is one of the main mechanisms for the much higher losses in POFs than in glass MMFs. However, the radiation mode coupling has little effect on the transmission bandwidth because the radiation loss due to volume scattering by the heterogeneities has no strong mode dependence. In this section, we revisit and develop a coupled power theory for understanding the intrinsic guided mode coupling in POFs.

The transverse electric fields in heterogeneous POFs can be approximately expressed as a superposition of the transverse guided mode fields of the ideal optical fiber without any inhomogeneities:

$$\boldsymbol{E}(x, y, z) = \sum_{i=1}^{N} c_i(z)\boldsymbol{E}_i(x, y)\exp(i\beta_i z), \tag{3.19}$$

where $\boldsymbol{E}_i$ and β_i are the transverse electric field vector and the propagation constant for mode i of the ideal optical fiber with N modes, respectively. In Equation 3.19, we ignore the contributions of the backward-propagating modes and radiation modes. For the guided mode fields, the following orthogonality condition is satisfied:

$$\frac{1}{\sqrt{2}}\int_{-\infty}^{\infty}\int_{-\infty}^{\infty} \boldsymbol{e}_z \cdot (\boldsymbol{E}_j \times \boldsymbol{H}_i^{\,*})dxdy = \delta_{ji}P, \tag{3.20}$$

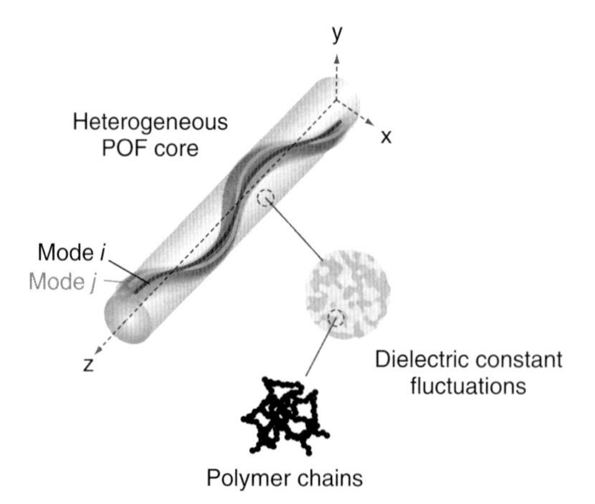

Figure 3.7 Random mode coupling due to microscopic heterogeneities.

where $\boldsymbol{e}_z$ is the unit vector along the fiber axis and P is the normalized coefficient. In a heterogeneous POF, the transverse electric field or the expansion coefficients of the guided light waves change with propagation by mode coupling of the heterogeneities, obeying the coupled mode equation

$$\frac{\partial c_i}{\partial z} = \sum_j K_{ij} c_j \exp[i(\beta_i - \beta_j)z] \tag{3.21}$$

with

$$K_{ij} = \frac{\omega}{4iP} \int_{-\infty}^{\infty} \int_{-\infty}^{\infty} \delta\varepsilon(x, y, z)(\boldsymbol{E}_i^* \cdot \boldsymbol{E}_j) dxdy. \tag{3.22}$$

Here, ω is the angular frequency of the guided light, and K_{ij} is the mode coupling coefficient. In the derivation of Equation 3.22, the contribution of the longitudinal field components is ignored in the weakly guiding approximation. As shown by Equation 3.22, the mode coupling coefficient depends on the exact distribution of the dielectric constant fluctuation. For random fluctuations of the microscopic heterogeneities, the ensemble-averaged power can be analyzed using the coupled power equation [28]

$$\frac{\partial P_i}{\partial z} + \tau_i \frac{\partial P_i}{\partial t} = -\alpha_i P_i + \sum_{j=1}^{N} h_{ij}(P_j - P_i), \tag{3.23}$$

where h_{ij} is the power coupling coefficient from mode j to mode i, τ_i is the group delay per unit length for mode i, and α_i is the attenuation coefficient for mode i. $P_{i(j)}$ is the ensemble-averaged mode power $<|c_i|^2>$ over slightly different optical fibers with statistically similar perturbations. In the original derivation of the coupled power equation, directional perturbations along the optical fiber axis (z-axis) are assumed because the fiber axis fluctuation and core diameter fluctuation can be the dominant sources of mode coupling in glass MMFs. On the other hand, the microscopic heterogeneities in POF cores are basically isotropic perturbations. Therefore, we use the generalized power coupling coefficients to take account of the perturbation isotropy [29]:

$$h_{ij} = \int_{-\infty}^{\infty} \langle K_{ij}(z) K_{ij}^*(z + \zeta) \rangle \exp[i\Delta\beta\zeta] d\zeta, \tag{3.24}$$

where $\Delta\beta$ $(= \beta_i - \beta_j)$ is the propagation constant difference. From Equations 3.17 and 3.24, we obtain the following power coupling coefficient for microscopic heterogeneities in POF core materials [30]:

$$h_{ij} = C_{ij} \iint |\boldsymbol{E}_i^* \cdot \boldsymbol{E}_j|^2 dxdy, \tag{3.25}$$

where

$$C_{ij} = \langle \delta\varepsilon^2 \rangle \frac{\omega^2 \pi^{\frac{3}{2}} D^3}{8P^2} \exp\left(-\frac{\Delta\beta^2 D^2}{4} \right). \tag{3.26}$$

The range of the integral on the right-hand side in Equation 3.25 is the core region. In the derivation of Equation 3.25, we assumed a shorter correlation length than the characteristic scale of the mode field variation, allowing the approximation of the Gaussian correlation function in the fiber cross section by the delta function product [31]. As shown in Equation 3.25, the power coupling coefficients are determined by the scattering characteristics of the materials and the field intensity overlap of the modes, which are given by C_{ij} and $\iint |\boldsymbol{E}_i^* \cdot \boldsymbol{E}_j|^2 dxdy$, respectively. Note that the coupling coefficient depends little on the propagation constant difference because $\exp(-\Delta\beta^2 D^2/4) \approx 1$ for the typical correlation length of the microscopic heterogeneities. This indicates that the mode coupling due to the heterogeneities should not be a diffusion process like the microbending-induced coupling, which occurs between nearest neighbor modes with the closest propagation constants.

3.2.4
Mode Coupling Mechanism

Using the developed coupled power theory, we can evaluate random mode coupling due to microscopic heterogeneities in PMMA-based GI POFs. Here, we analyze the parabolic refractive index profile whose guided modes can be classified into principal mode groups with almost the same group velocity, which helps in understanding the optical pulse transmission of the GI POF. The index profile is given by Equation 3.4, where $g = 2$ and $a = 25\ \mu\text{m}$. For typical PMMA cores doped with diphenyl sulfide, $n_1 = 1.503$ and $n_2 = 1.490$ for a laser source wavelength of 650 nm [32]. To calculate the power coupling coefficients using Equations 3.25 and 3.26, the propagation constants, mode fields, and group delays of the GI POF were calculated using finite element analysis of the scalar wave equation for the linearly polarized $\text{LP}_{m,l}$ modes, where m is the azimuthal mode number and l is the radial mode number. The GI POF has 147 LP modes, which can be classified into 24 principal mode groups. The highest order mode group was ignored in the analyses of the impulse response because the modes in the group can easily radiate from the fiber owing to their much lower power confinement in the core than other groups.

Figure 3.8 shows C_{ij} as a function of the correlation length for $\langle \delta\varepsilon_{\text{r}}^2 \rangle = 2.0 \times 10^{-8}$, where $\delta\varepsilon_{\text{r}}$ is the relative dielectric constant fluctuation. For guided mode coupling (closed circles), C_{ij} increases approximately with the cube of the correlation length, as mentioned in the previous subsection. The increase is attributed to the scattering profile change from isotropic scattering to forward scattering, which has more small-angle components corresponding to differences in the guided mode propagation constant. For the radiative mode coupling (dotted lines), the C_{ij} value depends strongly on the $|\Delta\beta|$ value, and the $|\Delta\beta|$ dependence is stronger for a longer correlation length. Figure 3.8 shows that the C_{ij} value decreases with an increase in the $|\Delta\beta|$ value or scattering angle. This indicates that a longer correlation length results in more forward scattering with fewer wide-angle components. Note also that all guided mode pairs have almost the same C_{ij} values. This indicates that the correlation lengths have little effect on the power transition pathway, which is determined by the mode dependence

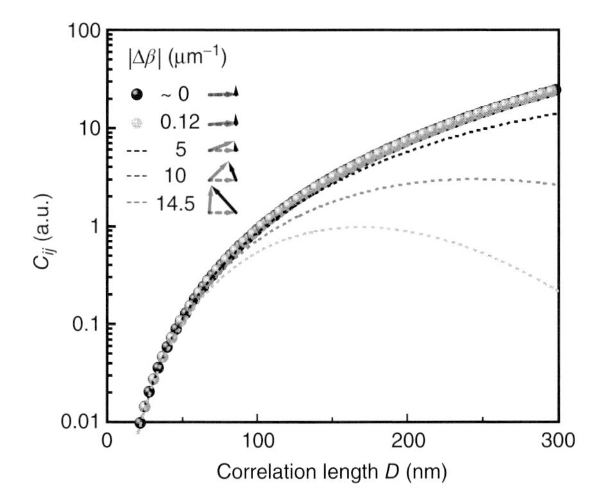

Figure 3.8 Coefficient C_{ij} as a function of correlation length D for several $|\Delta\beta|$ values, including the minimum and maximum values for guided mode pairs. Insets schematically show scattering vectors (black solid arrows) with propagation vectors of scattered (gray solid arrows) and incident (gray dotted arrows) light along the z-axis.

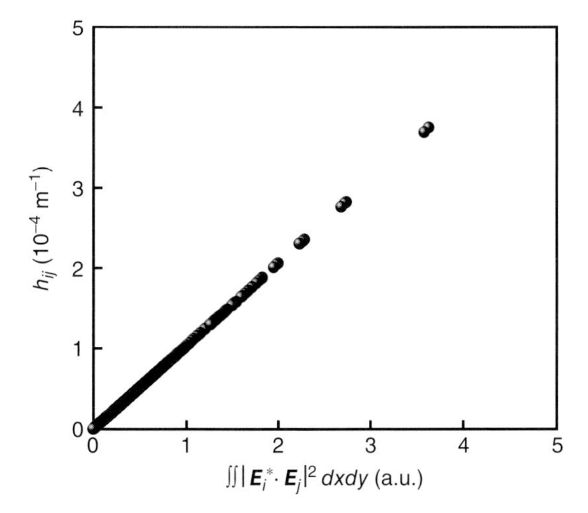

Figure 3.9 Power coupling coefficient as a function of $\iint |\boldsymbol{E}_i^* \cdot \boldsymbol{E}_j|^2 dxdy$.

of the coupling coefficient. Figure 3.9 shows the power coupling coefficient as a function of $\iint |\boldsymbol{E}_i^* \cdot \boldsymbol{E}_j|^2 dxdy$ for $D = 300$ nm. The coupling coefficient is approximately proportional to $\iint |\boldsymbol{E}_i^* \cdot \boldsymbol{E}_j|^2 dxdy$ because all the guided mode pairs have almost the same C_{ij} values. This indicates that the power transition pathway is predominantly determined by the mode field intensity overlaps, which depend on the structural parameters of the GI POFs.

As the parabolic index profile is not optimum for PMMA-based GI POFs, random mode couplings can affect pulse spreading owing to group delay averaging.

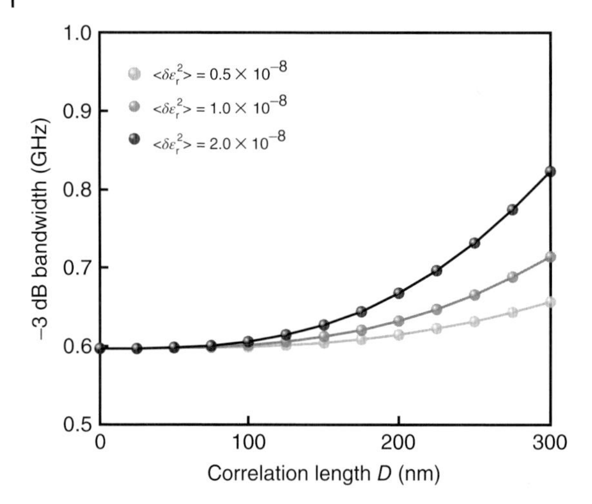

Figure 3.10 −3 dB bandwidth of 200-m GI POF as a function of correlation length for several $\langle \delta\varepsilon_r^2 \rangle$ values.

Using the power coupling coefficient h_{ij} and the modal delay per unit length τ_i, we investigate the optical pulse response of a GI POF at a wavelength of 650 nm in an overfilled launch condition with a homogeneous mode power distribution for all x-polarized LP modes using the numerical analysis of Equation 3.23 considering LP mode degeneracy. Here, only the x-polarized LP mode couplings are evaluated, assuming that the optical source is linearly polarized and that the GI POF has no structural or material anisotropies. In addition, the attenuation is not taken into account in the numerical calculation because it has no significant effect on the pulse response in the absence of mode-dependent attenuation. Actually, the radiative mode coupling or scattering loss due to microscopic heterogeneities, which is the origin of the attenuation, has no strong dependence on the mode number. Figure 3.10 shows the −3 dB bandwidth of a 200-m GI POF as a function of the correlation length for several $\langle \delta\varepsilon_r^2 \rangle$ values. The results show that the bandwidth can increase with the size and/or amplitude of the density fluctuations in the core polymers. This should be attributed to group delay averaging through random power transitions [28], whose transition rates are equally enhanced for all the mode pairs through changes in the scattering directions and/or efficiencies, as shown in Equations 3.25 and 3.26. This suggests that the optical fiber bandwidths can depend on the macroscopically observed GI profiles as well as on the microscopic material properties.

3.2.5
Efficient Group Delay Averaging

To further understand the intrinsic mode coupling, we investigated the optical pulse response of GI POFs in detail [33]. Figure 3.11a shows the power coupling coefficients for intrinsic mode coupling due to microscopic heterogeneities with

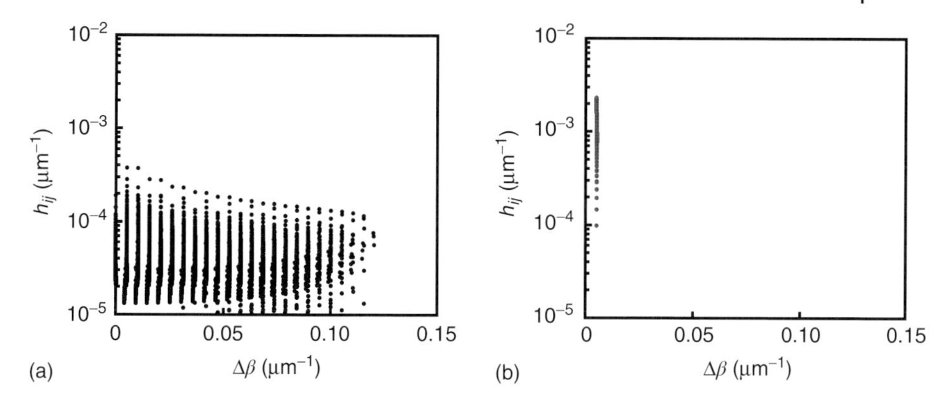

Figure 3.11 Power coupling coefficients h_{ij} as a function of propagation constant difference $\Delta\beta$ for mode coupling due to (a) microscopic heterogeneities and (b) microbending.

$D = 300$ nm and $\langle \delta\varepsilon_r^2 \rangle = 2.0 \times 10^{-8}$ as a function of the propagation constant difference of the coupled mode pair. It was confirmed that the $\Delta\beta$ values can be approximately classified into 24 discrete values corresponding to all the mode group pairs because of the parabolic index profile. Therefore, the result indicates that microscopic heterogeneities can induce random power transitions between all the group pairs. As shown in Figure 3.11b, we also calculated the coupling coefficient for microbending-induced coupling assuming that the bending curvature is Gaussian-correlated [34]. The standard deviation and correlation length of the curvature are 2×10^{-4}/mm and 1.0 mm, respectively, which are close to the reported values for glass MMFs. Microbending causes only coupling between the nearest neighbor groups, obeying the selection rules for mode coupling [21]. Note that the average coupling coefficient of 4.1×10^{-5}/m for microscopic heterogeneities is much smaller than the coefficient of 8.4×10^{-4}/m for microbending.

For the parabolic refractive index profiles, the optical pulse response of GI POFs can be understood by considering mode couplings between the principal mode groups. Figure 3.12a,b shows the average coupling coefficient H_{MN} between mode groups with the principal mode numbers M and N calculated for microscopic heterogeneities and microbending, respectively. Note that H_{11} and H_{22} have no values because mode groups 1 and 2 have only one guided mode each. The results show that forward scattering by microscopic heterogeneities can contribute to both intragroup ($M = N$) and intergroup ($M \neq N$) mode couplings, whereas only nearest neighbor couplings can be induced by microbending. Note also that intrinsic mode coupling due to microscopic heterogeneities is stronger for lower order mode groups owing to higher degrees of mode field intensity overlap. However, the coupling coefficients for microbending-induced coupling become larger for higher order group pairs, as shown in Figure 3.12b.

Figure 3.13a–c shows the output pulse waveform from GI POFs with microscopic heterogeneities, with microbending, and without any perturbations, calculated for different fiber lengths. The temporal pulse shape of the incident light is Gaussian with a full width at half-maximum of 83 ps. Owing to intermodal

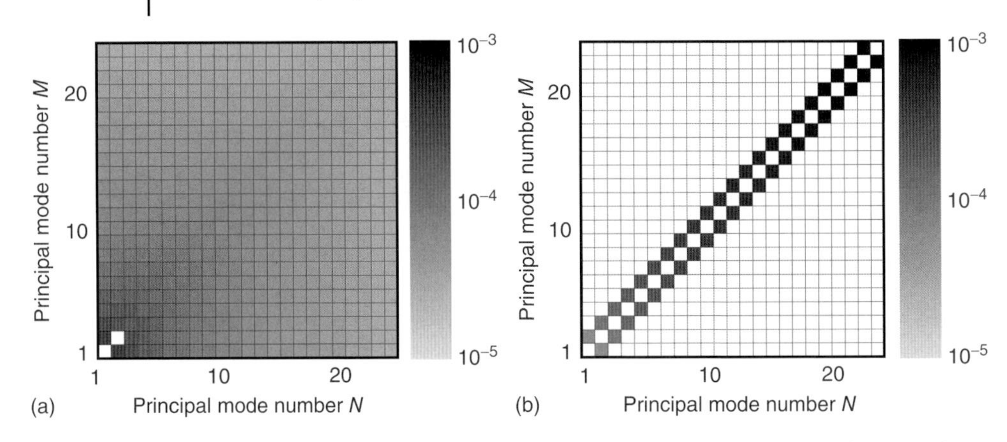

Figure 3.12 Average power coupling coefficients between mode groups with principal mode numbers M and N for random mode coupling due to (a) microscopic heterogeneities and (b) microbending.

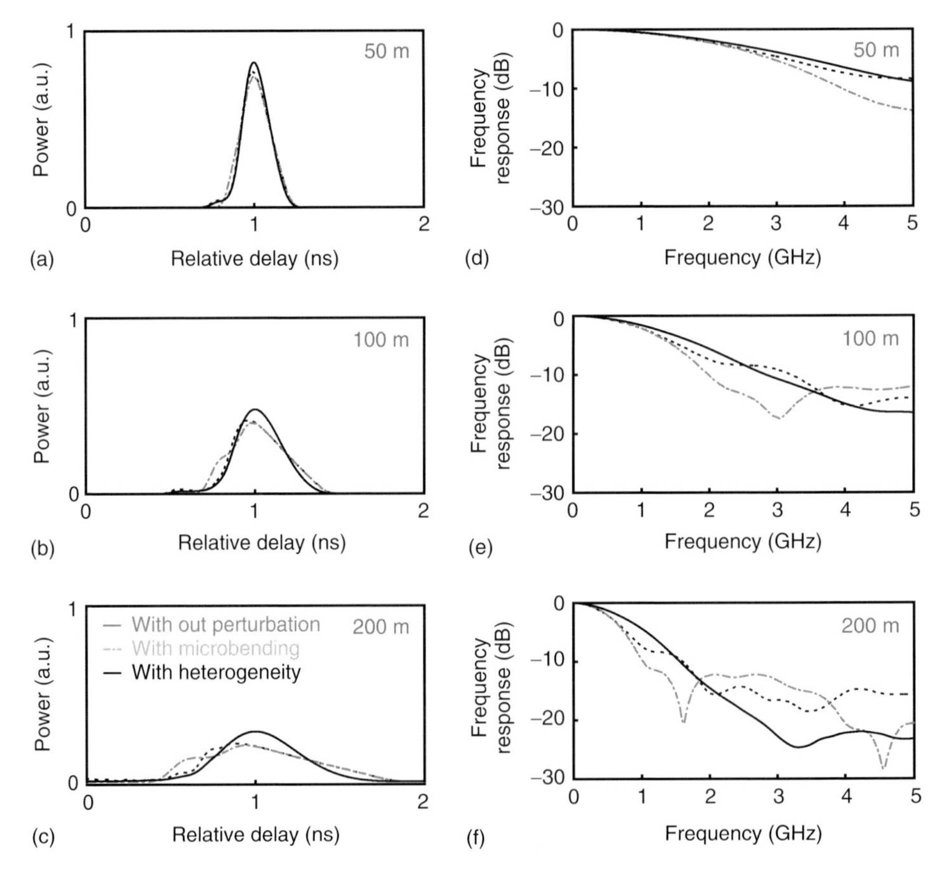

Figure 3.13 Output pulse waveform for lengths of (a) 50 m, (b) 100 m, and (c) 200 m, and relative frequency response for lengths of (d) 50 m, (e) 100 m, and (f) 200 m.

dispersion, the optical pulses are broadened with propagation in all the GI POFs. For GI POFS having a parabolic profile, higher order mode groups should have shorter group delays, which are approximately inversely proportional to the group order except for some groups near the cutoff. Thus, the output pulse waveform from the ideal GI POF without any perturbations exhibits an approximately linear decay with time in the pulse trailing edge (Figure 3.13c), which reflects the fact that a larger number of guide modes exist in higher order mode groups. The pulse waveforms are changed by perturbation through group delay averaging owing to random mode coupling. For the GI POFs with microbending, the mode coupling effect is pronounced only in the pulse buildup because the nearest neighbor coupling becomes stronger for higher order mode groups, as shown in Figure 3.13b. However, the microscopic heterogeneities affect the entire pulse waveform, and the pulse is broadened with little change in its pulse shape. Figure 3.13d–f shows the corresponding frequency response of GI POFs with microscopic heterogeneities, with microbending, and without any perturbations for the different fiber lengths. These results show that the random mode couplings increase the −3-dB bandwidth of the GI POF. Note that microscopic heterogeneities result in much stronger bandwidth enhancement than that due to microbending despite their much smaller average coupling coefficient, as mentioned above. This indicates that microscopic heterogeneities can efficiently average the group delays of all the guided modes because of the random power transition.

Figure 3.14 shows the pulse broadening as a function of fiber length for GI POFs with microscopic heterogeneities, with microbending, and without any perturbations. The pulse broadening was obtained by calculating the rms widths of the impulse responses. The ideal GI POF exhibits a linear increase in pulse

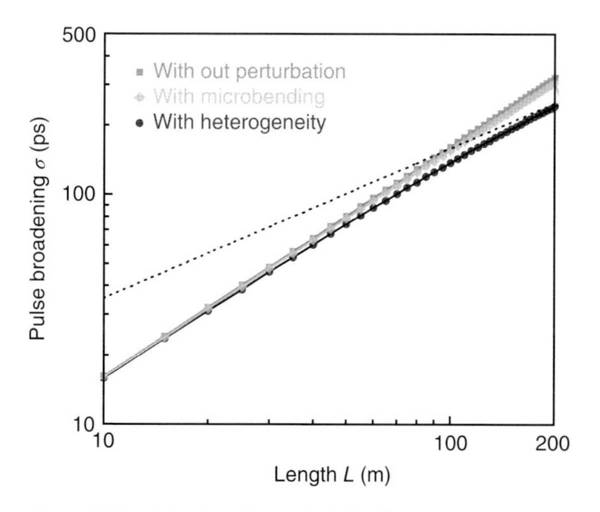

Figure 3.14 Pulse broadening in GI POFs with microscopic heterogeneities, with microbending, and without any perturbations. Dotted line is the fitted curve of $\sigma \propto L^{0.65}$ for equilibrium mode coupling of GI POF with microscopic heterogeneities.

broadening with increasing fiber length because of the linear length dependence of the maximum delay difference. In GI POFs with microscopic heterogeneities, intrinsic mode coupling can decrease the pulse broadening through group delay averaging. As shown in Figure 3.14, the pulse broadening is approximately proportional to $L^{0.65}$ for sufficiently long fibers, yielding equilibrium power mixing, whereas a linear dependence ($\sigma \propto L$) is observed for shorter lengths with little mode coupling effect. The resultant coupling length is ~100 m, at which the two asymptotes ($\sigma \propto L$ and $\sigma \propto L^{0.65}$) intersect. However, microbending has little effect on pulse broadening at the evaluated fiber lengths, and the resultant coupling length is much longer, although it has a much higher average coupling coefficient than microscopic heterogeneities.

3.3 Mode Coupling Effect in POFs

3.3.1 Radio-over-Fiber with GI POFs

RoF technology has attracted much interest for in-home and in-building networks [35–37], which can be enriched by transmission of various radio frequency signals over optical fibers. TV signal transmission over fiber allows simplified wiring than with the conventional coaxial and/or twist-pair cables in home networks. Moreover, distributed antenna systems can be realized in buildings by wireless and mobile signal transmission over fiber. For these short-reach applications, a practical RoF system with an SMF is too costly to apply. Therefore, MMF-based RoF systems have attracted considerable interest because of their much lower cost. However, MMF optical links have suffered from complex noise problems such as modal noise and reflection noise [38], especially in RoF systems whose modulation depths are much lower than those for on–off digital signal transmission. We recently encountered some noise reduction effects in multimode optical links with GI POFs, which are closely related to the intrinsic mode coupling due to microscopic heterogeneities [39, 40]. In the next subsection, we demonstrate that reflection noise can be significantly reduced in RoF systems by using microscopically heterogeneous GI POFs. This suggests that GI POFs can be used to introduce RoF systems in home and building networks.

3.3.2 Noise Reduction Effect in GI POFs

Reflection noise degrades the transmitted signal through laser intensity fluctuations due to optical feedback from some interfaces such as the fiber end faces and the photodiode (PD) [41–43]. This can be eliminated by using optical isolators and optical fibers with obliquely polished end faces. However, they cannot be applied to in-home and in-building networks because of their high installation cost.

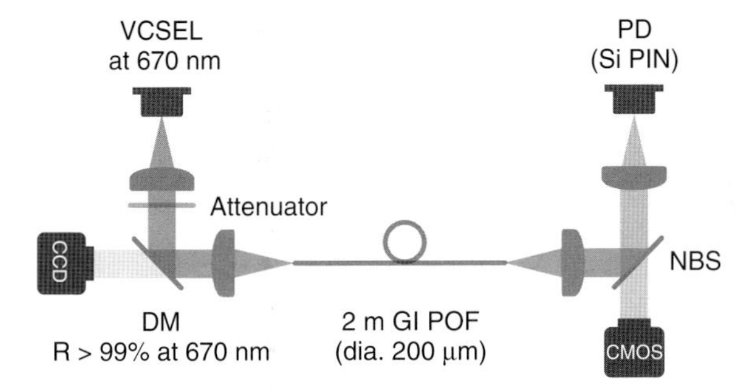

Figure 3.15 Experimental setup. PD: photodiode; DM: dichroic mirror; NBS: nonpolarization beam splitter.

Here we demonstrate that the reflection noise can be significantly reduced in RoF systems by using GI POFs with a microscopic heterogeneous core.

Figure 3.15 shows the experimental setup. The laser is a vertical-cavity surface-emitting laser (VCSEL) (Optowell, PM67-F1P1N) with continuous-wave oscillation at ~670 nm. The photodetector is a silicon PIN diode (New Focus, 1601-AC) with a −3-dB bandwidth of 1 GHz. We measured the noise-floor spectra of the MMF links with an unmodulated VCSEL, which was operated with a drive current of 3.0 mA. In this study, high-frequency intensity noise due to optical feedback from the fiber output face and the PD were mainly evaluated because the reflection noise can appear in the frequency bands for CATV and wireless signals. We did not evaluate the optical feedback from an MMF input face, which typically results in low-frequency intensity noise owing to the close distance between an MMF and a VCSEL in practical optical modules. For precise control of the launching conditions, we monitored a fiber near-end face irradiated with a laser beam using a charge-coupled device (CCD) camera. The PD irradiated with output beams from MMFs was also monitored to detect all of the output beams with the PD. This allowed the evaluation of the reflection noise without the effect of modal noise due to partial detection of the beam.

Figure 3.16 shows microscopic images of the input beams on fiber near-end faces and the output beams on the PD in an optical link with a reference fiber consisting of a 2-m glass GI MMF with a core diameter of 200 μm, an NA of 0.2, and an attenuation of 10 dB/km. Because the glass MMF has little mode coupling, the output mode power distribution is almost the same as the initial distribution for the launched modes, resulting in a near-field pattern in which the power is spatially localized around the core center for the center launching condition.

We developed a GI POF with an acrylic-polymer-based heterogeneous core, whose attenuation was ~200 dB/km at 670 nm. The attenuation produces a radiation loss of around 50 dB/km because of the intrinsic scattering due to microscopic core heterogeneities. The core diameter and NA were 200 μm and 0.3, respectively. Figure 3.17 shows microscopic images of the input beams on the fiber near-end faces and the output beams on the PD for the GI POF under the center launching condition. As shown in Figure 3.17b, the output beam pattern of the GI POF

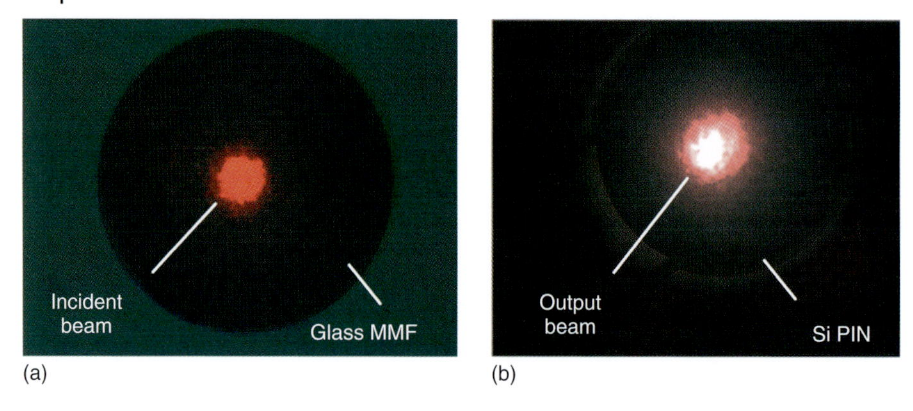

Figure 3.16 Microscopic images of (a) input beam on the fiber near-end faces and (b) output beam on Si PIN PD in optical link with glass MMF.

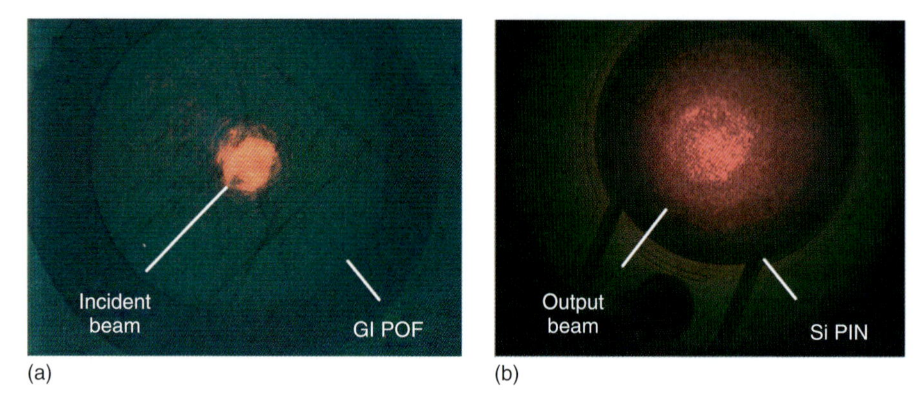

Figure 3.17 Microscopic images of (a) input beam on the fiber near-end faces and (b) output beam on Si PIN PD in optical link with a microscopically heterogeneous GI POF.

differs substantially from that of the glass MMF. In the GI POF, strong mode coupling results in a pronounced power transition from the launched modes to the other guided modes, whereas in glass MMFs the mode power distribution changes little from the initial one.

As shown in Figure 3.17b, the speckled pattern of the output beam indicates that intrinsic mode coupling results in a pronounced power transition from the launched modes to the other guided modes, even for a length of 2 m. The likely reason is that the mode coupling mechanism in the GI POF differs from the diffusion processes in glass MMFs. The random power transition can degrade the beam quality of the reflected light in the GI POF, affecting the optical feedback. In the following, the effects of the reflection noise in the GI POF and glass MMF are compared.

Figure 3.18 shows the noise floor spectra of optical links with the GI POF and the glass MMF. We also evaluated a glass MMF with an obliquely polished far-end face at 8°. For the glass MMF, the spectrum has peaks with an equal spacing

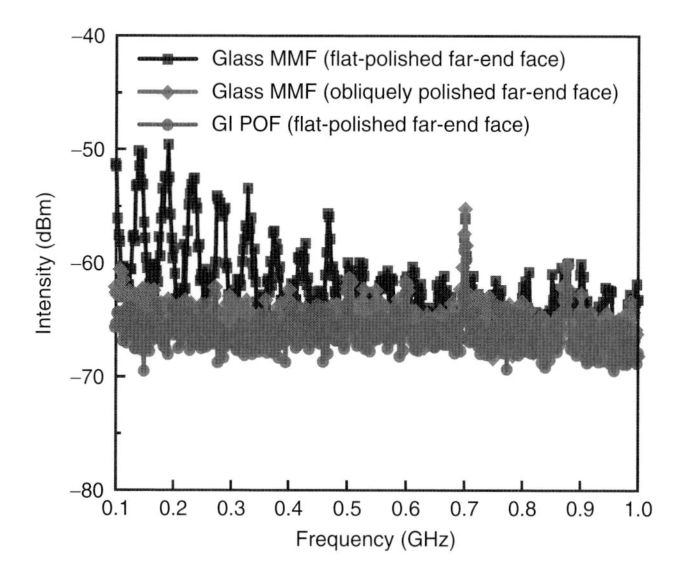

Figure 3.18 Noise floor spectra of optical links with 2-m GI POF and 2-m glass MMFs under center launching condition. Near-end faces of all the fibers are polished flat.

of ∼50 MHz, which corresponds to the round-trip frequency of the 2-m fiber. This suggests that the spikes are longitudinal mode beat notes of the fiber external cavity of the VCSEL [43]. On the other hand, spikes and reflection noise were not observed in the spectrum of the GI POF. Moreover, the floor levels of the GI POF were lower than those of the obliquely polished glass MMF, which can exhibit only reflection noise due to the PD. This suggests that the MMF links with the GI POF are affected little by the reflected light from both the end face and the PD.

We evaluated the influence of fiber attenuation on the reflection noise because the high attenuation in GI POFs can reduce the reflected light power or reflection noise. As shown in Figure 3.15, an attenuator consisting of an antireflection-coated neutral density filter was inserted to adjust the reflected power without changing the light path. Figure 3.19 shows the effect of 2- and 3-dB attenuators on the noise floor spectra of the glass MMF. Note that the GI POF attenuation of ∼0.4 dB is lower than value for the attenuators. Nevertheless, the GI POF has lower floor levels than the glass MMF with the attenuators. This indicates that the significant reduction in the reflection noise cannot be attributed to the higher attenuation of the GI POF compared to the glass MMFs.

To investigate the origin of the significant reduction in the reflection noise, the back-reflected beam patterns were microscopically observed using a higher power laser. Note that the microscopic images were captured with different CCD sensitivities for different fibers. Figure 3.20a shows a reflected beam pattern from the glass MMF with an obliquely polished far-end face. The Gaussian beam spot at the core center is the back-reflected beam from the near-end face, which corresponds to the incident beam pattern because the glass MMF has little back-reflected light

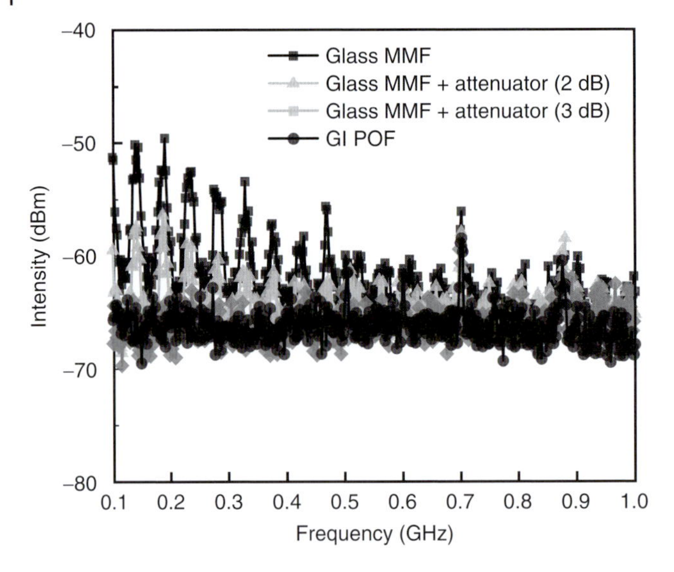

Figure 3.19 Noise floor spectra for GI POF, glass MMF, and glass MMF with 2- and 3-dB attenuators. All the fibers have flat-polished near-end and far-end faces.

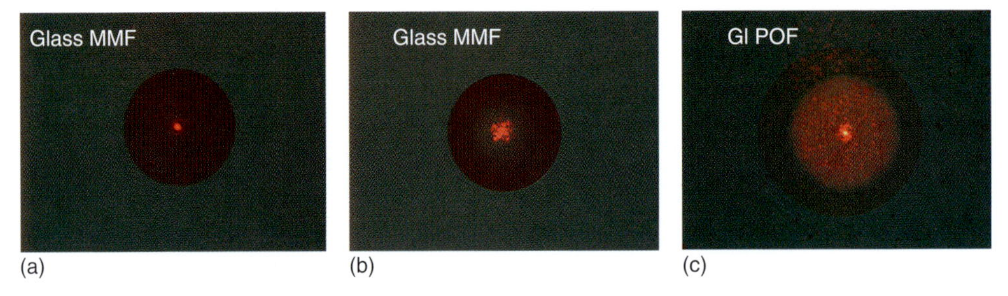

Figure 3.20 Microscopic images of back-reflected beam patterns from (a) glass MMF with obliquely polished far-end face, (b) glass MMF with flat-polished far-end face, and (c) GI POF with flat-polished far-end face under center launching conditions.

from its far-end face. For the glass MMF with a flat-polished far-end face, however, a speckled pattern in the back-reflected beam from the far-end face can also be observed, as shown in Figure 3.20b. Note that the reflected light speckle pattern has a mode power distribution similar to that of the initial mode launching condition. On the other hand, the GI POF exhibits reflected light speckled patterns filling the entire core, as shown in Figure 3.20c. This indicates that mode coupling results in pronounced power transitions from the launched modes to almost all the guided modes, even for a length of 2 m.

Figure 3.21 shows the reflected beam patterns from the fibers under an off-axis launching condition with an offset of 25 μm. For the GI POF, the reflected beam has a speckled pattern similar to that for the center launching condition (Figure 3.20c), whereas the reflected pattern reflects the initially launched mode

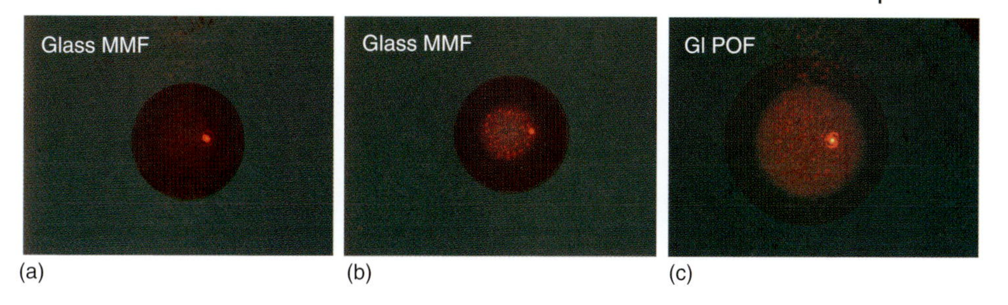

Figure 3.21 Microscopic images of back-reflected beam patterns from (a) glass MMF with obliquely polished far-end face, (b) glass MMF with flat-polished far-end face, and (c) GI POF with flat-polished far-end face under off-axis launching with an offset of 25 μm.

power distribution for the glass MMF. Therefore, the reflected beam's speckled pattern in the GI POF does not depend on the launching conditions. This suggests that the mode coupling effect almost reaches the steady state even for round-trip propagation in the 2-m GI POF. This strong mode coupling is likely due to the difference in the mode coupling mechanism of forward scattering in the GI POF from the diffusion processes of the glass MMFs. The random power transitions can degrade the spatial coherence of the reflected light in the GI POF. Therefore, the significant noise reduction under the center launching condition could be attributed mainly to the lowered self-coupling efficiency to the VCSEL cavity through the beam quality change.

References

1. Keiser, G. (2010) *Optical Fiber Communications*, Wiley Encyclopedia of Telecommunications, John Wiley & Sons, Inc..
2. Poole, C. and Wagner, R. (1986) Phenomenological approach to polarisation dispersion in long single-mode fibres. *Electron. Lett.*, **22**, 1029–1030.
3. Nowell, M.C., Cunningham, D.G., Hanson, D.C., and Kazovsky, L.G. (2000) Evaluation of Gb/s laser based fibre LAN links: review of the Gigabit Ethernet model. *Opt. Quantum Electron.*, **32**, 169–192.
4. Ziemann, O., Krauser, J., Zamzow, P.E., and Daum, W. (2008) *POF Handbook: Optical Short Range Transmission Systems*, 2nd edn, Springer-Verlag, Berlin.
5. Gloge, D. and Marcatili, E.A.J. (1973) Multimode theory of graded-core fibers. *Bell Syst. Tech. J.*, **52**, 1563–1578.
6. Olshansky, R. (1979) Propagation in glass optical waveguides. *Rev. Mod. Phys.*, **51** (2), 341–367.
7. Marcatili, E. (1977) Modal dispersion in optical fibers with arbitrary numerical aperture and profile dispersion. *Bell Syst. Tech. J.*, **56**, 49–63.
8. Croft, T.D., Ritter, J.E., and Bhagavatula, V.A. (1985) Low loss dispersion-shifted single-mode fiber manufactured by the OVD process. *J. Lightwave Technol.*, **3**, 931–934.
9. DiDomenico, M. Jr., (1972) Material dispersion in optical fiber waveguides. *Appl. Opt.*, **11**, 652–654.
10. Koike, Y. (1991) High-bandwidth graded-index polymer optical fibre. *Polymer*, **32**, 1737–1745.
11. Olshansky, R. and Keck, D.B. (1976) Pulse broadening in graded-index optical fibers. *Appl. Opt.*, **15**, 483–491.

12. Soudagar, M. and Wali, A. (1993) Pulse broadening in graded-index optical fibers: errata. *Appl. Opt.*, **32**, 6678.
13. Koike, Y. and Ishigure, T. (2006) High-bandwidth plastic optical fiber for fiber to the display. *J. Lightwave Technol.*, **24**, 4541–4553.
14. Ishigure, T., Koike, Y., and Fleming, J.W. (2000) Optimum index profile of the perfluorinated polymer-based GI polymer optical fiber and its dispersion properties. *J. Lightwave Technol.*, **18**, 178–184.
15. Polley, A., Gandhi, R.J., and Ralph, S.E. (2007) 40Gbps links using plastic optical fiber. Proceedings of Conference on Optical Fiber Communication/National Fiber Optic Engineers Conference, Anaheim, CA, 2007, p. OMR5.
16. Nuccio, S.R., Christen, L., Wu, X., Khaleghi, S., Yilmaz, O., Willner, A.E., and Koike, Y. (2008) Transmission of 40 Gb/s DPSK and OOK at 1.55 μm through 100 m of plastic optical fiber. Proceedings of European Conference and Exhibition on Optical Communication, Brussels, Belgium, September 21–25, 2008, p. We.2.A.4.
17. Yang, H., Lee, S.C.J., Tangdiongga, E., Okonkwo, C., van den Boom, H.P.A., Breyer, F., Randel, S., and Koonen, A.M.J. (2010) 47.4 Gb/s transmission over 100 m graded-index plastic optical fiber based on rate-adaptive discrete multitone modulation. *J. Lightwave Technol.*, **28** (4), 352–359.
18. Shi, R.F., Koeppen, C., Jiang, G., Wang, J., and Garito, A.F. (1997) Origin of high bandwidth performance of graded-index plastic optical fibers. *Appl. Phys. Lett.*, **71** (25), 3625–3627.
19. White, W., Reed, W.A., and Knudsen, E. (2003) Quantitative estimates of mode coupling and differential modal attenuation in perfluorinated graded-index plastic optical fiber. *J. Lightwave Technol.*, **21** (1), 111–121.
20. Polley, A. and Ralph, S.E. (2007) Mode coupling in plastic optical fiber enables 40-Gb/s performance. *IEEE Photonics Technol. Lett.*, **19** (16), 1254–1256.
21. Olshansky, R. (1975) Mode coupling effects in graded-index optical fibers. *Appl. Opt.*, **14** (4), 935–945.
22. Koike, Y., Tanio, N., and Ohtsuka, Y. (1989) Light scattering and heterogeneities in low-loss poly(methyl methacrylate) glasses. *Macromolecules*, **22** (3), 1367–1373.
23. Koike, Y., Matsuoka, S., and Bair, H.E. (1992) Origin of excess scattering in poly(methyl methacrylate) glasses. *Macromolecules*, **25** (18), 4807–4815.
24. Kerker, M. (1969) *The Scattering of Light and Other Electromagnetic Radiation*, Academic Press, New York.
25. Debye, P. and Bueche, A.M. (1949) Scattering by an inhomogeneous solid. *J. Appl. Phys.*, **20** (6), 518–525.
26. Debye, P., Anderson, H.R., and Brumberger, H. (1957) Scattering by an inhomogeneous solid. II. The correlation function and its application. *J. Appl. Phys.*, **28** (6), 679–683.
27. Born, M. and Wolf, E. (1999) *Principles of Optics: Electromagnetic Theory of Propagation, Interference and Diffraction of Light*, 7th edn, Cambridge University Press, London.
28. Marcuse, D. (1974) *Theory of Dielectric Optical Waveguides*, Academic Press.
29. Crosignani, B., Daino, B., and Di Porto, P. (1975) Statistical coupled equations in lossless optical fibers. *IEEE Trans. Microwave Theory Tech.*, **23** (5), 416–420.
30. Inoue, A., Sassa, T., Makino, K., Kondo, A., and Koike, Y. (2012) Intrinsic transmission bandwidths of graded-index plastic optical fibers. *Opt. Lett.*, **37** (13), 2583–2585.
31. Puente, N.P., Chaikina, E.I., Herath, S., and Yamilov, A. (2011) Fabrication, characterization, and theoretical analysis of controlled disorder in the core of optical fibers. *Appl. Opt.*, **50** (6), 802–810.
32. Koike, Y. and Koike, K. (2011) Progress in low-loss and high-bandwidth plastic optical fibers. *J. Polym. Sci. B*, **49** (1), 2–17.
33. Inoue, A., Sassa, R., Furukawa, T., Makino, K., Kondo, A., and Koike, Y. (2013) Efficient group delay averaging in graded-index plastic optical fiber with microscopic heterogeneous core. *Opt. Express*, **21** (14), 17379–17385.
34. Kitayama, K., Seikai, S., and Uchida, N. (1980) Impulse response prediction

based on experimental mode coupling coefficient in a 10-km-long graded-index fiber. *IEEE J. Quantum Electron.*, **16**, 356–362.

35. Sauer, M., Kobyakov, A., and George, J. (2007) Radio over fiber for picocellular network architectures. *J. Lightwave Technol.*, **25** (11), 3301–3320.
36. Gomes, N.J., Nkansah, A., and Wake, D. (2008) Radio-over-MMF techniques—Part I: RF to microwave frequency systems. *J. Lightwave Technol.*, **26** (15), 2388–2395.
37. Koonen, A.M.J. and García Larrodé, M. (2008) Radio-over-MMF techniques—Part II: microwave to millimeter-wave systems. *J. Lightwave Technol.*, **26** (15), 2396–2408.
38. Peterman, K. (1991) *Laser Diode Modulation and Noise*, Kluwer Academic Publishers, Dordrecht.
39. Inoue, A., Furukawa, R., Matsuura, M., and Koike, Y. (2014) Reflection noise reduction effect of graded-index plastic optical fiber in multimode fiber link. *Opt. Lett.*, **39** (12), 3662–3665.
40. Matsuura, M., Furukawa, R., Matsumoto, Y., Inoue, A., and Koike, Y. (2013) Evaluation of modal noise in graded-index silica and plastic optical fiber links for radio-over-multimode fiber systems. *Opt. Express*, **22** (6), 6562–6568.
41. Bae, J.W., Temkin, H., Swirhun, S.E., Quinn, W.E., Brusenbach, P., Parsons, C., Kim, M., and Uchida, T. (1993) Reflection noise in vertical cavity surface emitting lasers. *Appl. Phys. Lett.*, **63** (11), 1480–1482.
42. Hirota, O., Suematsu, Y., and Kwok, K. (1981) Properties of intensity noises of laser diodes due to reflected waves from single-mode optical fibers and its reduction. *IEEE J. Quantum Electron.*, **17** (6), 1014–1020.
43. Chen, Y.C. (1980) Noise characteristics of semiconductor laser diode coupled to short optical fibers. *Appl. Phys. Lett.*, **37** (7), 587–589.

4
Materials

Since the first POF (plastic optical fiber) was invented in the mid-1960s, considerable efforts have been made to develop materials in order to enhance POF's performance. The fundamental requirements for choosing or designing materials for POF fabrication are that the polymer should be (i) completely transparent, (ii) resistant to high temperatures, (iii) able to be drawn into a fiber, and (iv) mechanically flexible. Although the temperature requirement depends on where and how the POFs are utilized, in most cases it is preferred that their T_g should be above 80 °C. The above-mentioned four requirements are applicable irrespective of whether polymers are used as the core or for cladding. However, the core polymer of graded-index (GI) POFs should also possess the following requirements: (v) low refractive index, and (vi) low material dispersion. Though the GI profile can be fabricated by several methods, as will be discussed in Chapter 5, the most established method today is to diffuse a dopant from the core into the cladding. In such a polymer–dopant system, the biggest concern is that the dopant significantly decreases T_g because of the plasticization effect. Thus, it is preferable that the refractive index of the polymer is as low as possible so that a lower dopant concentration can be used to form a GI profile. The material dispersion is not important for step-index (SI) POFs because their bandwidth is almost dominated by modal dispersion. However, for GI POFs, this can be crucial in determining the transmission capacity of the fibers.

In this chapter, materials for fabricating POFs, especially their cores, are discussed. Section 4.1 describes the properties and problems of typical base polymer materials, namely poly(methyl methacrylate) (PMMA) and CYTOP®. In Sections 4.2 and 4.3, recent material developments that have been attempted to solve the difficulties of conventional polymers are detailed.

4.1
Representative Base Polymers of POFs

4.1.1
Poly(methyl methacrylate)

PMMA is a mass-produced, commercially available polymer that provides excellent resistance to both chemical and weather corrosion. The transmittance

Fundamentals of Plastic Optical Fibers, First Edition. Yasuhiro Koike.

of PMMA is the highest among general optical polymers; PMMA is known to transmit 93–94% of visible light and reflect the remainder. Since the first POF commercialized by Mitsubishi Rayon in 1975, most SI POFs have been manufactured using PMMA. They have been used extensively in industrial field buses for controlling process equipment in rugged manufacturing environments, and in automobiles to connect an increasing array of multimedia equipment. The increasing complexity of in-vehicle electronic systems in particular has led to PMMA-based SI POFs becoming indispensable to the automobile industry. Today, it is not uncommon to find 10 or 20 consumer electronic devices installed in cars, for example, DVD (blu-ray) players, navigation systems, telephones, Bluetooth interfaces, voice-recognition systems, high-end amplifiers, and TV tuners. To meet all the necessary requirements for data transfer among such devices, PMMA-based SI POFs, which offer up to several hundred megabits per second, have proven to be an ideal solution.

However, if bit rates of data transmission faster than 1 Gbps are required, attenuation becomes a problem in the PMMA fibers. Short-distance application systems have employed PMMA-based SI POFs as the transmission medium and red light-emitting diodes (LEDs) with a wavelength of 650 nm as the light source. As shown in Figure 2.10, 650 nm is in the low-loss window of PMMA. On the other hand, to realize gigabit communications, not only does the fiber have to be replaced with a GI POF but also an appropriate light source with a higher modulation bandwidth is necessary. At present, vertical-cavity surface-emitting lasers (VCSELs), which offer bit rates of up to 10 Gbps, are believed to be the most reasonable solution. VCSELs are a relatively new class of semiconductor lasers that are expected to become a key device in gigabit Ethernet, high-speed local-area networks, computer links, and optical interconnects. However, one problem remains even with VCSELs: their emission wavelengths are longer than 670 nm and are attenuated by the PMMA-based POFs to more than 200 dB/km because of the C–H stretching vibrational absorption loss. Thus, the transmission distance is severely limited.

As mentioned in Section 2.3.2, there have been several reports on perdeuterated PMMA (PMMA-d_8) as a base material for fabricating lower loss POFs. The replacement of hydrogen with deuterium in PMMA results in a considerable reduction in the C–H vibrational absorptions in the IR region and in its overtones in the visible to near-IR region. As a result, loss reductions to 20 and 63 dB/km at 670–680 nm for an SI POF [1] and a GI POF [2], respectively, were successfully achieved using PMMA-d_8 as the core base material. The refractive index profile of a GI POF was nearly optimized, and the −3-dB bandwidth was enhanced to 1.2 GHz over 300 m in an over-field-launch condition.

However, the humidity resistance of these fibers is a problem. Because the substitution of hydrogen with deuterium does not seriously affect most physical properties of the polymer, PMMA-d_8 exhibits a relatively high water absorption rate, up to 2 wt%, similar to PMMA. Thus, despite the fairly low attenuation over a wide range of wavelengths, PMMA-d_8-based POFs show high attenuations under

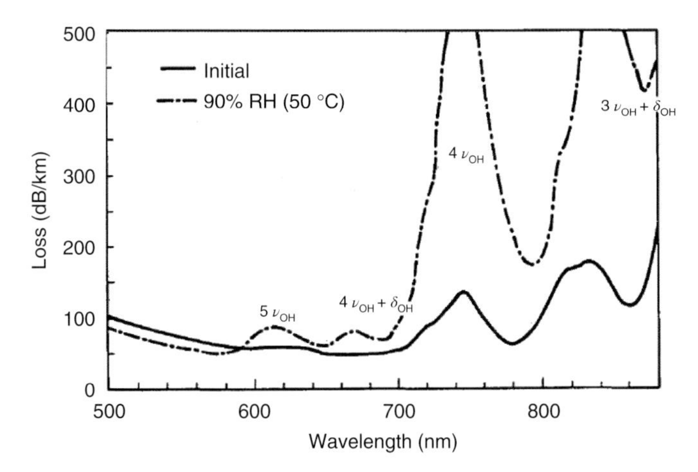

Figure 4.1 Attenuation spectra of PMMA-d$_8$-based SI POF before and after heavy humidity test at 90% RH and 50 °C.

high-humidity environments. Figure 4.1 shows the change in the attenuation spectrum of a PMMA-d$_8$-based SI POF before and after an accelerated moisture test at 90% RH and 50 °C [3]. For example, the two peaks of the broken line at 614 and 668 nm correspond to the fifth overtone of the O–H stretching vibrational absorption ($5\nu_{OH}$) and a combination of the fourth overtone ($4\nu_{OH}$) and O–H bending vibrational absorptions (δ_{OH}), respectively. At longer wavelengths, the absorption peaks due to the O–H bonds of absorbed water and C–D bonds of the polymer overlap. In addition to this disadvantage, the monomer and polymer are too expensive for common POFs. Thus, PMMA-d$_8$ is not currently considered as a potential POF material.

4.1.2 Perfluorinated Polymer, CYTOP®

Because the high attenuation of PMMA is caused by overtones of the C–H vibrational absorption, the most effective method for obtaining a low-loss POF is to substitute all the hydrogen atoms with heavier atoms such as fluorine, which helps to avoid the water absorption problem mentioned above. The replacement of hydrogen with fluorine also improves the thermal stability, chemical resistance, and electrical properties of POFs. The size of the fluorine atom allows the formation of a uniform and continuous sheath around the carbon–carbon bonds and protects them from attack, thus imparting heightened chemical resistance and stability to the molecule.

However, compared to the large number of radically polymerizable hydrocarbon monomers, only a few classes of perfluorinated monomers can homopolymerize under normal conditions via a free-radical mechanism. The most typical example is tetrafluoroethylene (TFE), developed by DuPont in 1938. As is well known, poly(tetrafluoroethylene) is opaque white, despite the

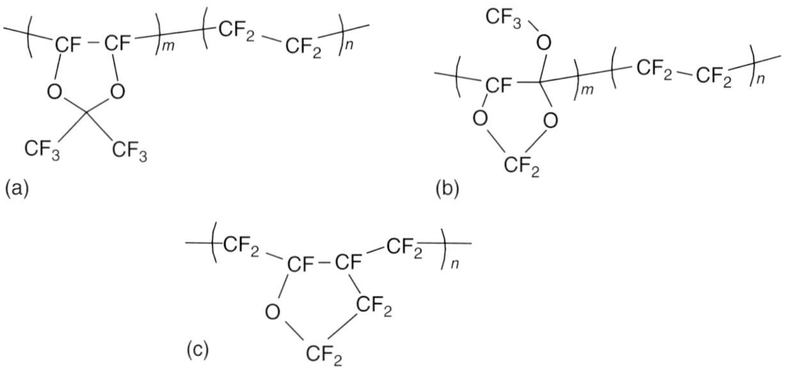

Figure 4.2 Chemical structures of (a) Teflon AF, (b) Hyflon AD, and (c) CYTOP.

lack of C–H bonds. TFE polymerizes linearly without branching, which gives rise to a virtually perfect chain structure and rather high molecular weights. The chains have minimal interactions and crystallize to form a nearly 100% crystalline structure. Other perfluorinated resins also have more or less similar properties and easily form at least partially crystalline structures. Hence, light is scattered at the boundaries between the amorphous and crystalline phases, causing haziness. Before the 1980s, all industrial fluoropolymers were semicrystalline materials.

To avoid the formation of the crystalline phase, introduction of aliphatic rings into the main chain, which causes it to become twisted, is an effective method. The most well-known amorphous perfluorinated polymers are Teflon® AF, Hyflon® AD, and CYTOP, developed by DuPont, Solvay Chemicals, and AGC, respectively. The chemical structures of these three polymers are shown in Figure 4.2. They have excellent clarity, solubility in some fluorinated solvents, thermal and chemical durability, low water absorption, and low dielectric properties. In particular, their high transparencies arise from the cyclic structures that exist in the polymer main chains. Teflon AF is a copolymer of 2,2-bis(trifluoromethyl)-4,5-difluoro-1,3-dioxole, which possesses a cyclic structure in its monomer unit, and TFE, while Hyflon AD is a copolymer of perfluoro-2,2,4-trifluoro-5-trifluoromethoxy-1,3-dioxole and TFE. CYTOP is a homopolymer of perfluoro(1-butenyl vinyl ether) (PFBVE), whose cyclopolymerization yields cyclic structures (five- and six-membered rings) on the polymer backbone. The synthesis procedure of PFBVE is shown in Scheme 4.1 [4]. At present, only CYTOP is utilized as a base material for POFs.

The first CYTOP-based GI POF (Lucina™), produced by the preform-drawing method, was commercialized by AGC in 2000. The attenuation spectrum has already been shown in Figure 2.10 along with that of the PMMA-based GI POF. As can be seen in Figure 4.2, CYTOP molecules consist solely of C–C, C–F, and C–O bonds. The wavelengths of the fundamental stretching vibrations of these bonds are considerably longer than those of the C–H bond, and therefore the vibrational absorption losses of CYTOP in the visible to near-IR region are negligibly small.

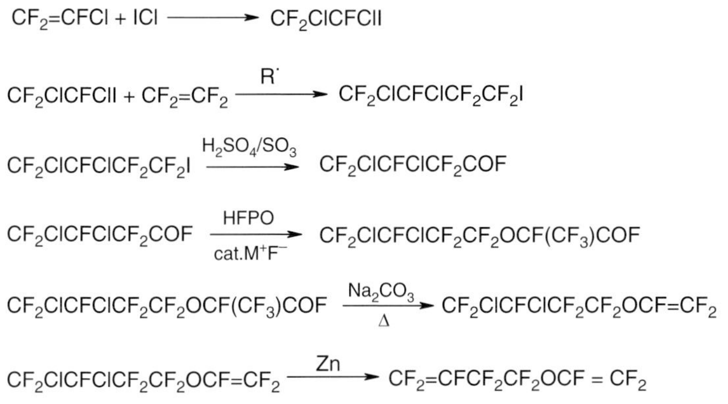

$$CF_2{=}CFCl + ICl \longrightarrow CF_2ClCFClI$$

$$CF_2ClCFClI + CF_2{=}CF_2 \xrightarrow{R^\cdot} CF_2ClCFClCF_2CF_2I$$

$$CF_2ClCFClCF_2CF_2I \xrightarrow{H_2SO_4/SO_3} CF_2ClCFClCF_2COF$$

$$CF_2ClCFClCF_2COF \xrightarrow[\text{cat.}M^+F^-]{HFPO} CF_2ClCFClCF_2CF_2OCF(CF_3)COF$$

$$CF_2ClCFClCF_2CF_2OCF(CF_3)COF \xrightarrow[\Delta]{Na_2CO_3} CF_2ClCFClCF_2CF_2OCF{=}CF_2$$

$$CF_2ClCFClCF_2CF_2OCF{=}CF_2 \xrightarrow{Zn} CF_2{=}CFCF_2CF_2OCF = CF_2$$

Scheme 4.1 Synthesis of PFBVE.

The excellent low-loss characteristics of CYTOP-based GI POFs are sometimes far beyond the requirements for short-range networks. However, the truly unique feature of these fibers is the low material dispersion derived from the low dielectric constant. This has been detailed in Section 3.1. The calculation clearly shows that CYTOP-based GI POFs may have a higher bandwidth than multimode GOFs (glass optical fibers); this was demonstrated in 1999 when AGC and Bell Laboratories reported an experimental transmission of 11 Gbps over 100 m using a CYTOP-based GI POF [5]. In 2008, a group at the Georgia Institute of Technology and a collaboration between the University of Southern California and Keio University separately set a new record of 40 Gbps [6, 7]. Subsequently, a group at the Technical University of Eindhoven established a transmission of 47.4 Gbps in 2010 [8]. In 2012, a collaboration between NEC Laboratories and the University of Florida established a 112-Gbps transmission over 100 m [9]. These highly significant results demonstrate that GI POFs can achieve higher bandwidths than multimode GOFs.

In 2010, AGC released another CYTOP-based GI POF called Fontex™. By employing a double cladding structure (a thin layer with a considerably lower refractive index placed around the first cladding), the bending loss was further reduced while the high-speed capacity was maintained, thereby enabling various wiring designs. In addition, the continuous fabrication of GI POFs was established by introducing the coextrusion process.

4.2 Partially Halogenated Polymers

4.2.1 Polymethacrylate Derivatives

There is no doubt that CYTOP is by far the best available material for fabricating GI POFs. CYTOP-based GI POFs continue to break records for the lowest attenuation and highest data transmission speed. However, despite their impressive performance, CYTOP-based GI POFs have not been widely accepted, mainly because of the prohibitive costs. CYTOP is a homopolymer fabricated by the cyclopolymerization of PFBVE, which is prepared by a complex process outlined in Scheme 4.1. The manufacturing costs are quite high, and CYTOP resin typically sells for several dollars per gram, which makes it one of the most expensive synthetic polymers. In fact, many of the actual intended purposes and potential applications of POFs do not require the high specifications of CYTOP. A low-cost fiber capable of a 1-Gbps transmission over ~30 m at the light source wavelength (670–680 nm) is currently in high demand. However, this transmission cannot be achieved with PMMA-based GI POFs because of high attenuation; at the same time, it is not necessary to use CYTOP. Thus, the partially fluorinated or chlorinated polymethacrylate derivatives discussed in this section have been studied as possible alternatives.

Recently, three partially halogenated methacrylates, shown in Figure 4.3, have been intensively investigated as core base materials for fabricating GI POFs: poly(2,2,2-trifluoroethyl methacrylate) (poly(TFEMA)) [10]; poly(2,2,2-trichloroethyl methacrylate) (poly(TClEMA)) [11–13]; and MMA-*co*-pentafluorophenyl methacrylate (PFPhMA) [14]. To form the refractive index profiles, poly(TFEMA) was doped with benzyl benzoate, and the other two methacrylates were doped with diphenyl sulfide (DPS). All the materials are commercially available; the monomers are slightly more expensive than MMA but considerably cheaper than PFBVE. Figure 4.4 shows the attenuation spectra of GI POFs based on these polymers compared to that of the PMMA-based GI POF in Figure 2.10. The poly(TFEMA)-based GI preform-drawn fiber was prepared by interfacial-gel polymerization, and the other two fibers were prepared by the rod-in-tube method. Further details of the fabrication procedure are given in the literature [10–14]. The purpose of employing partially halogenated polymers is to decrease the fiber attenuation; the most important factor here is how many C–H bonds exist per unit volume. This can be calculated using the density of the polymer, molecular weight of a monomer unit, and number of C–H bonds per monomer unit. The C–H bond concentrations of poly(TFEMA) and poly(TClEMA) are 64% and 51%, respectively, of the corresponding value

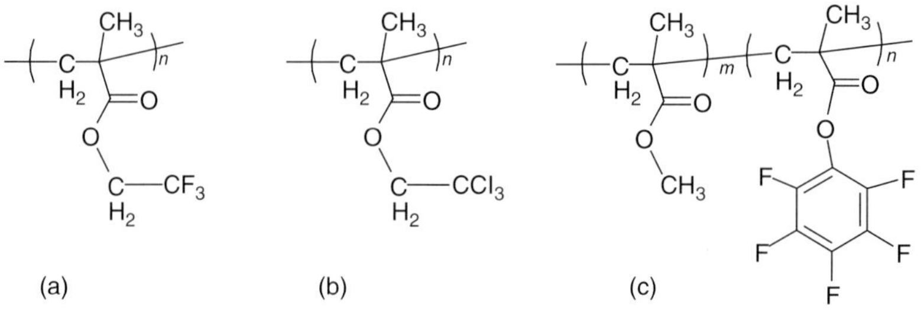

Figure 4.3 Chemical structures of (a) poly(TFEMA), (b) poly(TClEMA), and (c) MMA-*co*-PFPhMA.

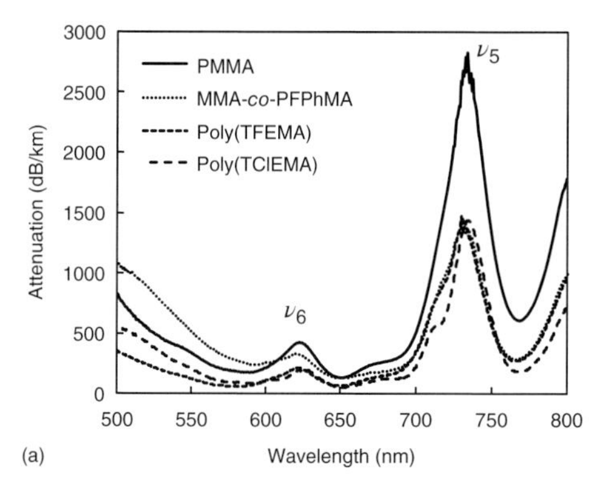

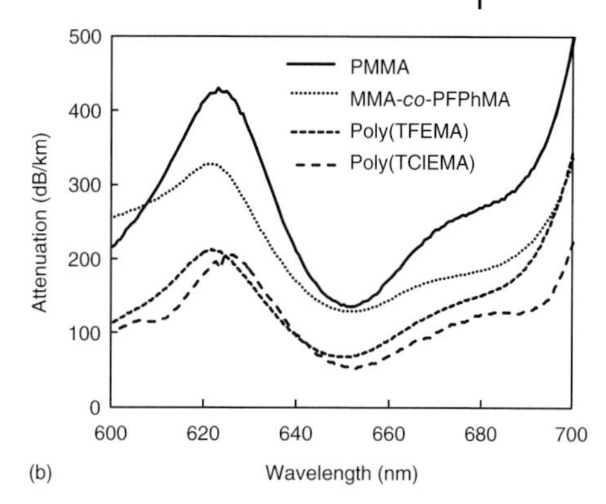

Figure 4.4 Attenuation spectra of GI POFs based on PMMA, poly(TFEMA), poly(TClEMA), and MMA-*co*-PFPhMA (65/35 mol%) at (a) 500–800 nm and (b) 600–700 nm. ν represents C–H vibrational absorptions of PMMA. The subscript number is the vibrational quantum number.

of PMMA. For a copolymer composition of MMA/PFPhMA = 65/35 mol%, the C–H bond concentration is 68%. These concentrations roughly reflect the change in the peak intensities of the fifth and sixth C–H overtones from those of PMMA. Because of the considerable reduction in the C–H vibrational absorption losses, lower attenuations – less than 200 dB/km – have been achieved for source wavelengths of 670–680 nm.

The low attenuation of the copolymer-based GI POF is particularly noteworthy. Copolymers generally tend to have high attenuation because the dielectric fluctuations resulting from the copolymerization of monomers with different reactivities can cause nonnegligible light scattering losses. This usually far exceeds the value expected from the intrinsic light scattering of each homopolymer. As discussed in Section 2.2.2, the light scattering from large heterogeneous structures $V_{\mathrm{V2}}^{\mathrm{iso}}$ shows an angular dependence, defined as follows:

$$V_{\mathrm{V2}}^{\mathrm{iso}} = \frac{8\pi^3\langle\eta^2\rangle D^3}{\lambda_0^4(1 + k^2s^2D^2)^2}, \tag{4.1}$$

$$k = \frac{2\pi}{\lambda}, \tag{4.2}$$

$$s = 2\sin\left(\frac{\theta}{2}\right). \tag{4.3}$$

Here, $\langle\eta^2\rangle$ is the mean square average of the fluctuations of all the dielectric constants, D is the correlation length, and θ is the scattering angle relative to the incident direction. When a two-phase model is assumed such as that described in Figure 4.5, the correlation length a, which is a measure of the degree of heterogeneity, is given as follows [15]:

$$a = \frac{4V}{S}V_{\mathrm{A}}V_{\mathrm{B}}, \tag{4.4}$$

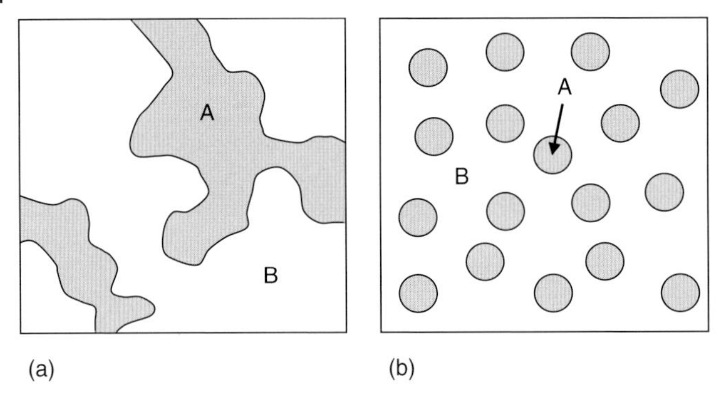

Figure 4.5 Two-phase models of heterogeneous structures. The correlation length D is (a) long and (b) short.

where S is the total contact area of the A and B phases with different dielectric constants, V is the total volume, and V_A and V_B are the volume fractions of the A and B phases ($V_A + V_B = 1$), respectively. Equations 4.1–4.4 indicate that V_{V2}^{iso} can be reduced in two ways. The first is to increase the contact area S between the A and B phases. By increasing S, the correlation length a becomes small, and V_{V2}^{iso} decreases; consequently, the ideal random or alternating copolymers are more transparent. The other way is to minimize the mean square average of the fluctuations of all dielectric constants $\langle \eta^2 \rangle$. If $\langle \eta^2 \rangle$ is zero, V_{V2}^{iso} is also zero, regardless of the monomer reactivity ratios. The copolymer of MMA and PFPhMA is such an example. Because the dielectric constant is almost equal to the square of the refractive index, the value of $\langle \eta^2 \rangle$ can be roughly estimated from the difference in the refractive indices of each homopolymer. In the copolymer case, the refractive indices are almost identical (PMMA: $n_D = 1.4914$; poly(PFPhMA): $n_D = 1.4873$), and thus V_{V2}^{iso} is negligible.

This is an extremely useful method for obtaining highly transparent copolymers. Under actual-use conditions, the transmission properties and the environmental resistance characteristics, such as the heat resistance, weatherability, and chemical resistance, must be considered; the mechanical properties are also important for installation requirements. However, identifying homopolymers that meet all the physical properties while maintaining the complete transparency required for optical communications is a difficult task. Copolymerization is a general method of modifying the physical properties in order to meet specific needs; however, it can result in high light scattering losses. Indeed, the number of ideal random copolymer and alternating copolymer combinations is severely limited. Furthermore, an accurate prediction of the monomer reactivity ratio is also quite difficult without performing the reaction. However, this method, which inhibits the excess scattering loss by adjusting the refractive index of each component, is more realistic and relatively easy. The refractive index decreases by introducing fluorine, whereas it increases with the introduction of chlorine, sulfone, phenyl, and so on. Using this method, several copolymer systems have been recently proposed as novel GI POF base materials [16–18].

In 2010, Sekisui Chemical commercialized a GI POF called *GINOVER* based on the studies of poly(TClEMA). While the polymer had some disadvantages – it depolymerized during the fabrication process and was considerably brittle – these problems were resolved by copolymerizing TClEMA with a small amount of *N*-cyclohexyl maleimide (cHMI) [12, 13]. Continuous fabrication of this GI POF was also recently established by the coextrusion process. The most distinctive feature of this fiber is its high temperature resistance. The attenuation of the copolymer-based GI POF, which is 132 dB/km at 660 nm, does not change after heating the fiber at 100 °C for over 2000 h. To the best of our knowledge, this is the first GI POF capable of maintaining low attenuation at 100 °C for such a long period.

4.2.2
Polystyrene Derivatives

Polystyrene (PSt), another universal optical polymer, is one of the most widely used plastics today. In terms of its mechanical properties and chemical resistance, PSt is slightly inferior to PMMA, but is very attractive as a core base material and is also relatively inexpensive. PSt has been continuously and extensively studied for use in SI POFs since the earliest days of POF development. The first PSt-based low-loss SI POF had an attenuation value of 114 dB/km at 670 nm [19]. This is the exact wavelength needed for gigabit data transmissions over GI POFs. PSt is a long-chain hydrocarbon, and its attenuation in the visible to near-IR region is also dominated by overtones of C–H vibrational absorption, as is the case with PMMA. What makes such low attenuation possible? The styrene unit is composed of aliphatic and aromatic C–H bonds that resonate at different wavelengths. The absorption wavelengths of the aliphatic C–H bonds corresponding to the fifth, sixth, and seventh overtones are 758, 646, and 562 nm, respectively, whereas the same overtones of the aromatic C–H bonds appear at 714, 608, and 532 nm, respectively. In addition, the positions of the aliphatic C–H overtones from the main chain are shifted slightly to longer wavelengths compared to those for PMMA because of an induction effect from the benzene ring. As a result, the emission wavelength of a VCSEL (670–680 nm) is located at the lowest attenuation window, which is in the center of the fifth aromatic and sixth aliphatic C–H overtones.

In 2012, the first PSt-based GI POF was reported [20]. The fiber was obtained by preform-drawing and rod-in-tube methods. Its attenuation spectrum is shown in Figure 4.6 alongside that of the PMMA-based GI POF. The *n*th overtones of the aliphatic and aromatic C–H bonds are labeled as ν_n and ν_n', respectively. The attenuation of the PSt-based GI POF is 166–193 dB/km at 670–680 nm, which is significantly lower than the attenuation of the PMMA-based GI POF in the same region (~240–270 dB/km).

In spite of this advantage, the possibility of PSt-based GI POFs was not investigated until recently because the refractive index of PSt is as high as 1.59. While this is advantageous for SI POFs because there are various possible cladding

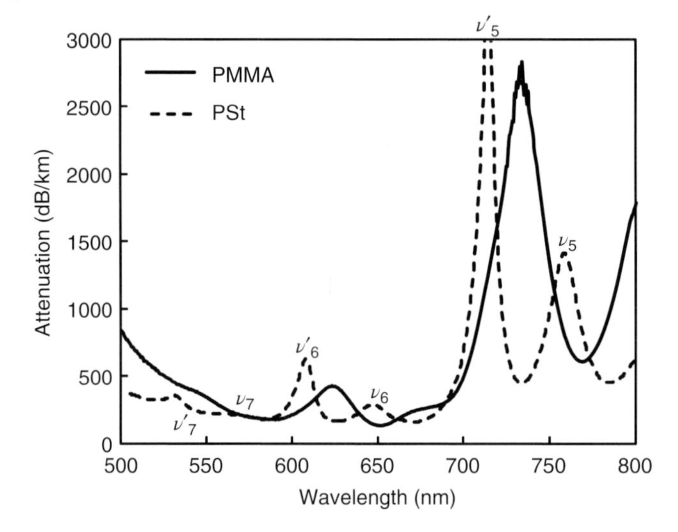

Figure 4.6 Attenuation spectra of PSt- and PMMA-based GI POFs. ν and ν′ are the aliphatic and aromatic C–H vibrational absorptions of PSt. The subscript number is the vibrational quantum number.

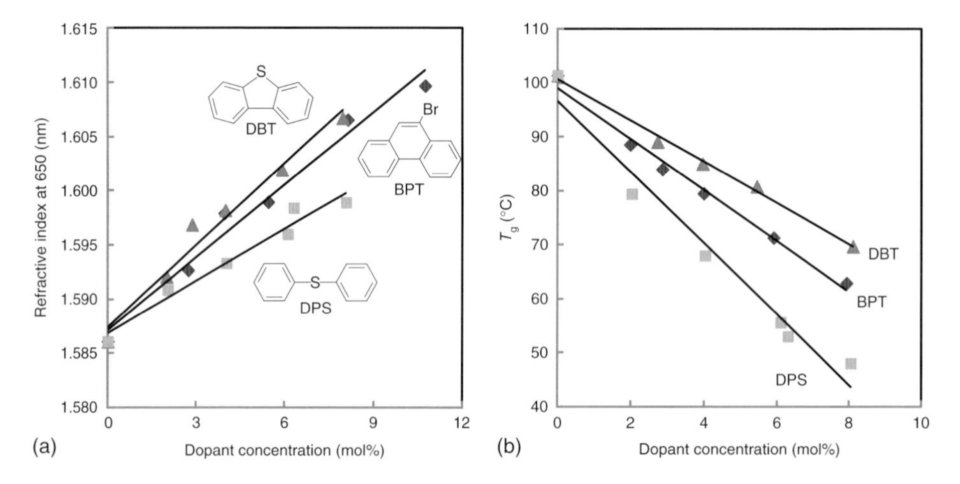

Figure 4.7 (a) Dopant concentrations and refractive indices of doped PSts and (b) dopant concentrations and T_g values of doped PSts.

polymers, in the case of GI POFs a higher refractive index means that a higher dopant concentration is necessary. In other words, T_g of the doped core polymer is further decreased because of the plasticization effect. Figure 4.7 shows the changes in the refractive index at 650 nm and T_g of doped PSt. These three dopants, DPS, dibenzothiophene (DBT), and 9-bromophenanthrene (BPT), are common for acrylic GI POFs. To design a PSt-based GI POF with NA 0.2, the refractive index of the core should be 1.5986. In the case of DPS, 6.3 mol% is needed, and T_g of the doped polymer is only 55 °C. While DBT and BPT are better dopants, their T_g values are still approximately 80 °C.

Introducing fluorine substituents is effective in lowering the refractive index, and recently various partially fluorinated derivatives of PSt have been studied [21–23]. Partially halogenated styrene monomers are generally prepared via dehydration of the corresponding alcohols, which are synthesized by via Grignard reactions with acetaldehyde. In a typical procedure, the Grignard reagents are prepared from an appropriate bromide compound and magnesium metal in anhydrous tetrahydrofuran (THF). After gradually adding acetaldehyde, the reaction mixture is treated with diluted hydrochloric acid. The resulting alcohols are subjected to distillation under vacuum. The monomers are obtained as colorless liquids on eliminating water from the alcohols using phosphorus pentoxide, followed by fractional distillation under vacuum. The obtained monomer is further purified by column distillation, and the polymer is obtained by free-radical polymerization using a typical thermal initiator. The synthesis procedure of the monomer is briefly summarized in Scheme 4.2.

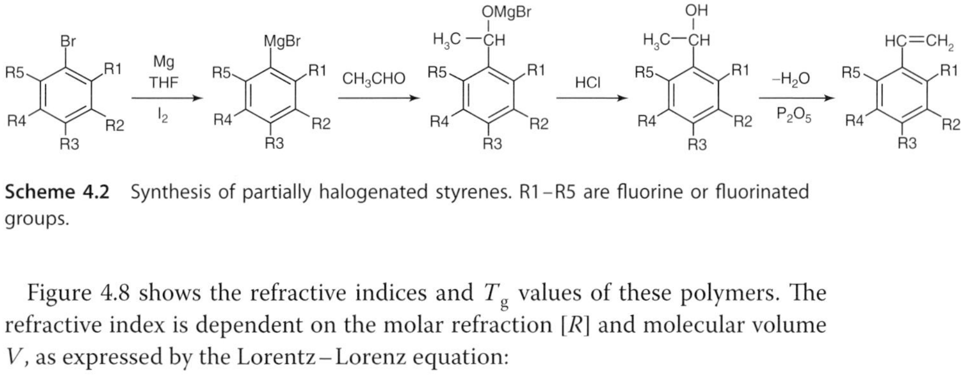

Scheme 4.2 Synthesis of partially halogenated styrenes. R1–R5 are fluorine or fluorinated groups.

Figure 4.8 shows the refractive indices and T_g values of these polymers. The refractive index is dependent on the molar refraction $[R]$ and molecular volume V, as expressed by the Lorentz–Lorenz equation:

$$n = \sqrt{\left(2\frac{[R]}{V} + 1\right) \Big/ \left(1 - \frac{[R]}{V}\right)}. \tag{4.5}$$

Because the fluorine substituents result in a smaller $[R]/V$ value than for hydrocarbons, the refractive index is lowered roughly in accordance with the proportion of fluorine atoms contained in each monomer unit. Figure 4.8 also shows that the refractive index is not significantly affected by the substitution site. This indicates that the density of the polymer, or, more specifically, the ratio of fluorine substituents per unit volume, does not significantly vary according to their position.

However, the polymer T_g is strongly affected by both the size and position of the substituent. When all the hydrogen atoms in the benzene ring are substituted with fluorine, that is, poly(2,3,4,5,6-pentafluorostyrene), T_g is almost the same as that of PSt. However, if hydrogen atoms at the ortho or para positions are substituted with CF_3 groups, then the polymer shows higher T_g values because such bulky substituents reduce the segmental mobility owing to the steric hindrance.

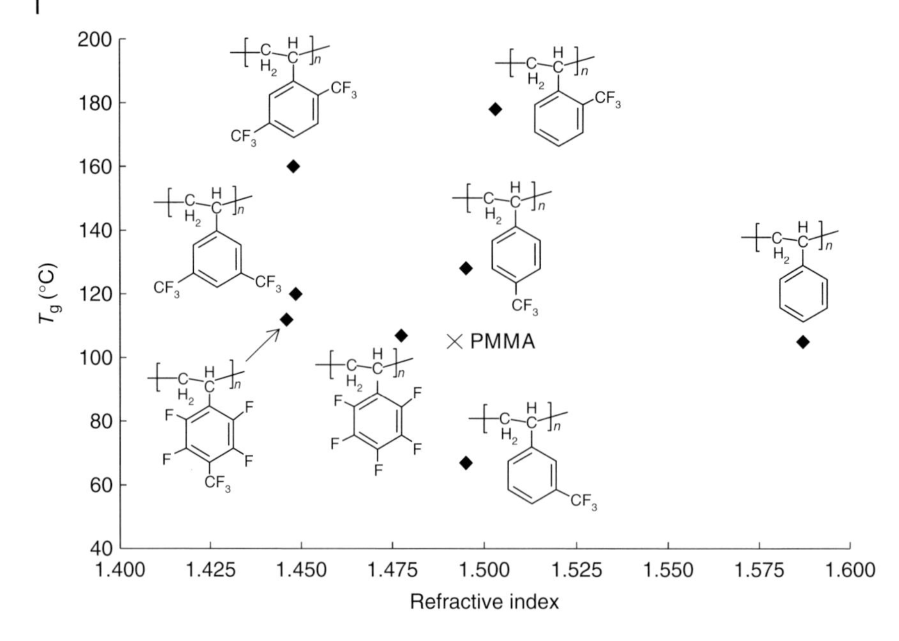

Figure 4.8 Refractive indices and T_g values of partially fluorinated derivatives of PSt.

In particular, a substituent at the ortho position, which is the nearest position to the main chain, is most efficient in enhancing T_g. Currently, a novel GI POF based on poly(2-trifluoromethyl styrene), which has an extremely high T_g of 178 °C, is under investigation.

4.3 Perfluoropolymers

4.3.1 Perfluorinated Polydioxolane Derivatives

The recent studies of partially halogenated polymers described in the previous sections were undertaken to obtain GI POFs with sufficiently low attenuation at specific light source wavelengths at a relatively low cost. These polymers can be categorized as intermediate materials between PMMA and CYTOP on the basis of their performance and material cost, both of which meet the current demand.

At the same time, considerable effort has been devoted to develop new perfluoropolymers as a substitute for CYTOP. CYTOP has excellent properties in many aspects, whereas the preparative methods are complex and expensive, as shown in Scheme 4.1, and the T_g is relatively low (~108 °C). In particular, several substituted perfluoro-2-methylene-1,3-dioxolanes, shown in Figure 4.9, have been extensively studied.

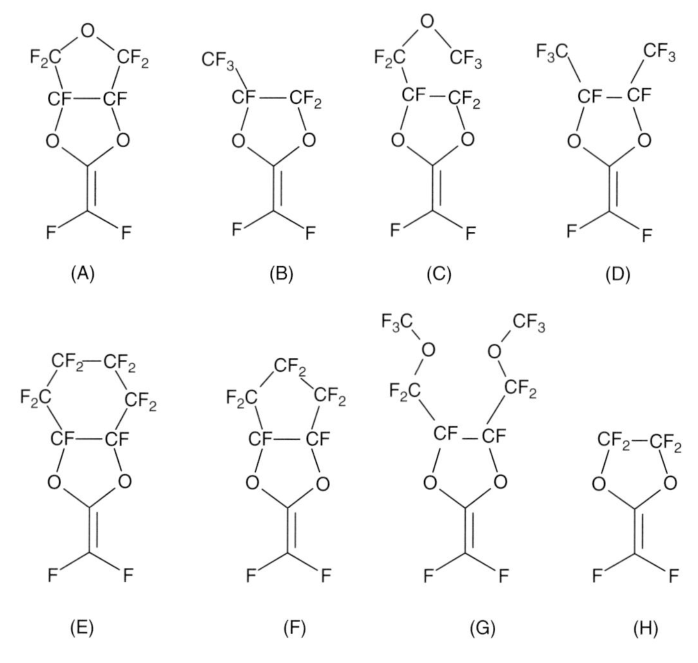

Figure 4.9 Chemical structures of perfluoro-2-methylene-1,3-dioxolanes.

The first example in this class is perfluoro-2-methylene-4-methyl-1,3-dioxolane (monomer B in Figure 4.9), which was prepared by DuPont in 1967 using perfluoropyruvyl fluoride prepared from hexafluoropropylene oxide [24]. Using a similar method, perfluoro-2-methylene-4,5-dimethyl-1,3-dioxolane (monomer D in Figure 4.9) was synthesized by Asahi Glass in 1993 [25].

One of the most efficient and simplest routes for the fabrication of these perfluorodioxolane monomers and polymers is shown in Scheme 4.3 [26–30]. Commercially available methyl pyruvate and diols are used as the starting materials. The condensation product is then directly fluorinated in a fluorinated solvent such as Fluorinert™ FC-75 (3 M) with a F_2/N_2 mixture. After flushing the system with nitrogen gas for 1 h, fluorine gas diluted to 20% with nitrogen is blown into the reaction mixture at a flow rate of 240 l/h. The reaction is carried out at approximately 25 °C for over 20 h. After the reaction is complete, the mixture is neutralized with aqueous potassium hydroxide. The crude monomer is isolated upon decarboxylation of the potassium salt and then purified by fractional distillation. The fluorination process is the key step in the synthesis. The yield is as high as 75% with careful control of the fluorination temperature, and it depends on the structure of the hydrocarbon precursor. Recently, over 90% yield of the fluorination of similar compounds has been reported [31]. The polymerization of these monomers is generally carried out in bulk at 60–80 °C for 24–35 h using perfluorodibenzoyl peroxide as the initiator [32]. The conversion yields are 70–80%, and the polymers obtained are colorless and transparent.

The polymers are purified by precipitation from their solutions in hexafluorobenzene in chloroform.

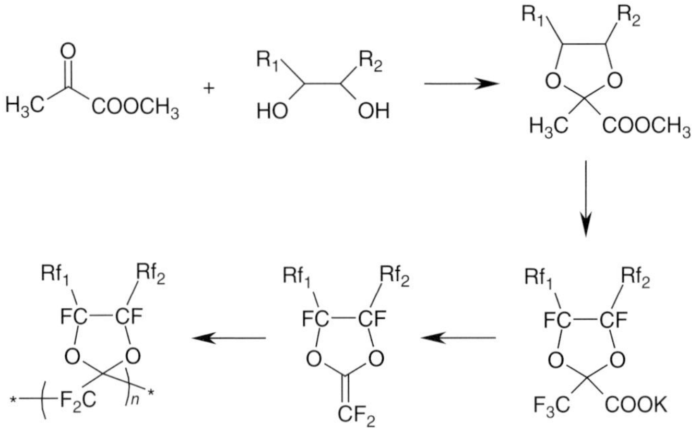

Scheme 4.3 Typical synthetic route for poly(perfluoro-2-methylene-1,3-dioxolane).

The decomposition of perfluorodibenzoyl peroxide occurs at 60–80 °C by a homolytic mechanism, resulting in the formation of a pentafluorobenzyl structural unit at the polymer terminus (Scheme 4.4) [31, 32]. Thus, poly(perfluoro-2-methylene-1,3-dioxolanes) are generally thermally stable, and their decomposition temperature (T_d) in air is >300 °C.

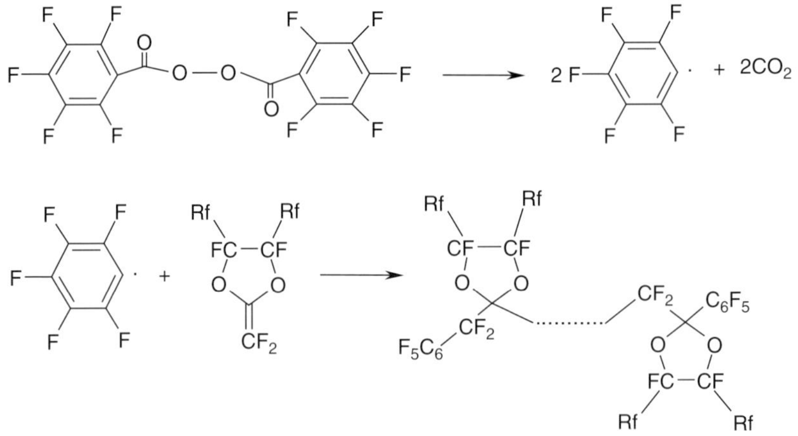

Scheme 4.4 Decomposition of initiator and structure of perfluoropolydioxolane polymer.

The molecular weight of the polymers can be controlled by the concentrations of the free-radical initiator and/or chain-transfer agents such as carbon tetrabromide. Typical results of the preparation of poly(perfluoro-2-methylene-4-methyl-1,3-dioxolane) using different concentrations of the initiator and chain-transfer agent are reported in [27].

Table 4.1 Properties of perfluoro-2-methylene-1,3-dioxolane monomers and their homopolymers.

Monomer	A	B	C	D	E	F	G	H
Polymerization rate ($\times 10^4$ mol/l·s)[a]	1.66	1.56	1.40	0.15	0.25	0.18	1.02	1.63
Polymer T_g (°C)	168	135	101	165	165	185	93	110
Polymer $RI_{632\ nm}$	1.3570	1.3310	1.3328	1.3280	1.3420	1.3460	1.3520	1.3443

a) In FC113, (monomer) = 1.6 mol/l, (perfluorodibenzoyl peroxide) = 0.05 mol/l, 41 °C.

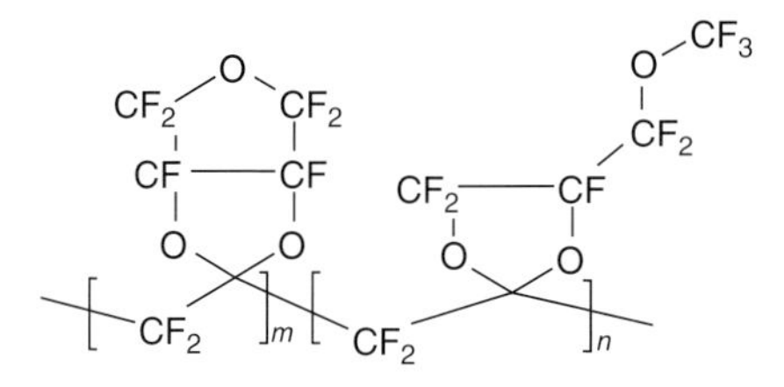

Figure 4.10 Chemical structure of a copolymer of perfluorodioxolanes (monomers A and C of Figure 4.9).

The physical properties of these monomers and their homopolymers are given in Table 4.1. These polymers, except the polymer of monomer H, are completely amorphous and exhibit extraordinary optical transmittance from the UV to the near-IR region. They are chemically and thermally stable. The polymers are soluble in fluorinated solvents such as hexafluorobenzene and Fluorinert F75. The polymer of monomer H was semicrystalline, and the melting point of the crystalline part is 230 °C. However, when the crystalline polymer powder was heated above the melting temperature and pressed, the prepared film became amorphous and flexible and did not recrystallize. The films were insoluble in any solvent, including fluorinated solvents [30, 33]. The 4- and 5-dimethyl substituted 1,3-dioxolane (Figure 4.9: D) has cis and trans isomers, with 27% and 73% proportions, respectively [34]. The T_g of these dioxolane polymers highly depends on the substituents at the fourth and fifth positions. The T_g of dimethyl and cyclic substituted polymers (Figure 4.9: A, D, E, and F) is relatively high (165 – 185 °C) because of the sterically crowded resultant polymers. On the other hand, the T_g of dialkylether-substituted polymers (Figure 4.9: G) is considerably lower (93 °C). The low T_g value is rationalized by the pendant group effect because the $–CF_2OCF_3$ group could have a larger free volume than CF_3 and cyclic substituted compounds.

The most remarkable observation in the series of dioxolane polymers is that some derivatives with relatively bulky substituents show excellent T_g values of up to 185 °C. Considering that the operating temperature of CYTOP-based GI POFs is limited to 70 °C, these high T_g values are highly attractive. However, such high T_g amorphous polymers tend to be brittle, and these perfluorodioxolane polymers are no exception. Thus, various methods for improving their mechanical properties and fiber preparation have been investigated. The following sections present examples of how their properties can be modified.

4.3.2 Copolymers of Dioxolane Monomers

Perfluoro-2-methylene-1,3-dioxolane monomers can be copolymerized with each other to modify the physical properties of the polymers. The refractive index and T_g depend on the copolymer composition. The copolymers are readily prepared in solution and in bulk. For example, the copolymerization reactivity ratios of monomers A and C (Figure 4.10) are $r_A = 0.97$ and $r_C = 0.85$ [35]. The data show that this copolymerization yields nearly ideal random copolymers. Figure 4.11 shows the change in T_g as a function of the copolymer composition. The copolymers have only one T_g, which increases from 110 to 165 °C as the mole fraction of monomer A increases. The copolymer films prepared by casting are flexible and tough and have a high optical transparency.

4.3.3 Copolymers of Perfluoromethylene Dioxolanes and Fluorovinyl Monomers

Perfluoro-2-methylene-1,3-dioxolane monomers can be also copolymerized with various commercially available fluorovinyl monomers. Perfluoro-3-methylene-2,4-dioxabicyclo[3.3.0]octane (Figure 4.9: F) was copolymerized with chlorotrifluoroethylene (CTFE), perfluoropropyl vinyl ether, perfluoromethyl vinyl ether,

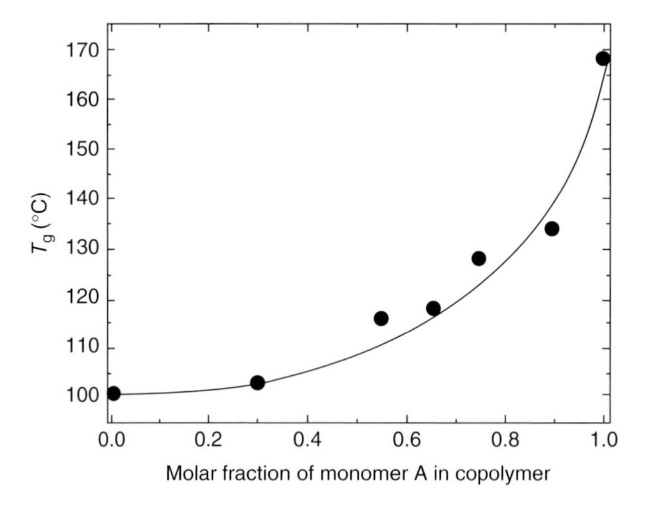

Figure 4.11 Dependence of T_g on copolymer composition (monomers A and C of Figure 4.9).

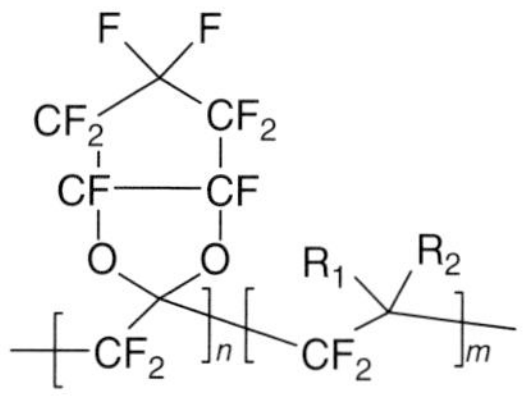

R_1: F, Cl, H, OCF_3, or OC_3H_7
R_2: F

Figure 4.12 Structure of copolymers based on perfluoro-3-methylene-2,4-dioxabicyclo[3.3.0]octane (monomer F) with fluorovinyl monomers.

Table 4.2 Copolymers of monomer F (perfluoro-3-methylene-2,4-dioxabicyclo[3.3.0]octane) with various fluorinated vinyl monomers (M2).

Monomer M_2	Monomer in feed (mol%)		Copolymer composition (mol%, EA)[a]		Copolymer composition (mol%, NMR)[b]		T_g (°C)	T_d (°C)	RI (633 nm)
	F	M_2	F	M_2	F	M_2			
$F_2C{=}CFCl$	12	88	16	84	18	82	—	—	—
$F_2C{=}CFCl$	20	80	35	65	37	63	84	322	—
$F_2C{=}CFCl$	40	60	62	38	62	39	110	330	1.3745
$F_2C{=}CFCl$	56	44	80	20	81	19	120	330	—
$F_2C{=}CFCl$	70	30	87	13	90	10	145	340	—
$F_2C{=}CFOC_3F_7$	65	35	87	13	91	9	144	342	1.3298
$F_2C{=}CFOCF_3$	21	79	51	49	60	40	154	359	1.3419
$F_2C{=}CH_2$	14	86	38	62	42	58	108	370	—
$F_2C{=}CH_2$	35	65	58	42	62	38	138	356	1.3517

a) The composition was determined by elemental analysis.
b) The composition was determined by NMR.

and vinylidene fluoride [36] using a free-radical initiator such as perfluorodibenzoyl peroxide or *tert*-butyl peroxypivalate in bulk or in solution, respectively (Figure 4.12). The typical results of these copolymers are shown in Table 4.2. The content of monomer F in the copolymer produced is considerably higher than that in the feed, indicating that the reactivity of the vinyl monomers is considerably lower than that of the perfluorinated dioxolane monomers. The reactivity ratios of CTFE and monomer F are $r_{CTFE} = 0.74$ and $r_F = 3.64$. The copolymers obtained are soluble in fluorinated solvents such as hexafluorobenzene and Fluorinert. The T_g of the copolymers is in the range 84–145 °C, and the copolymers are thermally stable (T_d: 320–370 °C). The copolymer films are flexible and transparent with a low refractive index (1.332–1.375 at 633 nm).

References

1. Kaino, T., Jinguji, K., and Nara, S. (1983) Low loss poly(methyl methacrylate-d8) core optical fibers. *Appl. Phys. Lett.*, **42** (7), 567–569.
2. Kondo, A., Ishigure, T., and Koike, Y. (2005) Fabrication process and optical properties of perdeuterated graded-index polymer optical fiber. *J. Lightwave Technol.*, **23** (8), 2443–2448.
3. Kaino, T. (1985) Influence of water absorption on plastic optical fibers. *Appl. Opt.*, **24** (23), 4192–4195.
4. Sugiyama, N. (1997) in *Modern Fluoropolymers: High Performance Polymers for Diverse Applications* (ed J. Scheirs), John Wiley & Sons, Ltd, Chichester, pp. 541–555.
5. Giaretta, G., White, W., Wegmueller, M., Yelamarty, R.V., and Onishi, T. (1999) 11 Gb/sec data transmission through 100 of perfluorinated graded-index polymer optical fiber. Technical Digest of Optical Fiber Communication Conference and Exhibit, San Diego, CA, February 21–26, 1999, p. PD14-1-3.
6. Polley, A. and Ralph, S.E. (2008) 100 m, 40 Gb/s plastic optical fiber link. Proceeding of Conference on Optical Fiber Communication/National Fiber Optic Engineers Conference, San Diego, CA, February 24–28, 2008, p. OWB2.
7. Nuccio, S.R., Christen, L., Wu, X., Khaleghi, S., Yilmaz, O., Willner, A.E., and Koike, Y. (2008) Transmission of 40 Gb/s DPSK and OOK at 1.55 μm through 100 m of plastic optical fiber. Proceeding of 34th European Conference on Optical Communication, Brussels, Belgium, September 21–25, 2008, p. We.2.A.4.
8. Yang, H., Lee, S.C.J., Tangdiongga, E., Okonkwo, C., van den Boom, H.P.A., Breyer, F., Randel, S., and Koonen, A.M.J. (2010) 47.4 Gb/s transmission over 100 m graded-index plastic optical fiber based on rate-adaptive discrete multimode modulation. *J. Lightwave Technol.*, **28** (4), 352–359.
9. Shao, Y., Cao, R., Huang, Y.-K., Ji, P.N., and Zhang, S. (2012) 112-Gb/s transmission over 100 m of graded-index plastic optical fiber for optical data center applications. Proceeding of Conference on Optical Fiber Communication/National Fiber Optic Engineers Conference, Los Angeles, CA, March 4–8, 2012, p. OW3J.
10. Koike, K. and Koike, Y. (2009) Design of low-loss graded-index plastic optical fiber based on partially fluorinated methacrylate polymer. *J. Lightwave Technol.*, **27** (1), 41–46.
11. Asai, M., Inuzuka, Y., Koike, K., Takahashi, S., and Koike, Y. (2011) High-bandwidth graded-index plastic optical fiber with low-attenuation, high-bending ability, and high-thermal stability for home-networks. *J. Lightwave Technol.*, **29** (11), 1620–1626.
12. Nakao, R., Kondo, A., and Koike, Y. (2012) Fabrication of high glass transition temperature graded-index plastic optical fiber: part 1-material preparation and characterizations. *J. Lightwave Technol.*, **30** (2), 247–251.
13. Nakao, R., Kondo, A., and Koike, Y. (2012) Fabrication of high glass transition temperature graded-index plastic optical fiber: part 2-fiber fabrication and characterizations. *J. Lightwave Technol.*, **30** (7), 969–973.
14. Koike, K., Kado, T., Satoh, Z., Okamoto, Y., and Koike, Y. (2010) Optical and thermal properties of methyl methacrylate and pentafluorophenyl methacrylate copolymer: design of copolymers for low-loss optical fibers for gigabit in-home communications. *Polymer*, **51** (6), 1377–1385.
15. Debye, P., Anderson, H.R., and Brumberger, H. (1957) Scattering by an inhomogeneous solid II. The correlation function and its application. *J. Appl. Phys.*, **28** (6), 679–683.
16. Koike, K., Mikes, F., Koike, Y., and Okamoto, Y. (2008) Design and synthesis of graded index plastic optical fibers by copolymeric. *Polym. Adv. Technol.*, **19** (6), 516–520.
17. Koike, K., Mikes, F., Okamoto, Y., and Koike, Y. (2009) Design, synthesis, and characterization of a partially chlorinated acrylic copolymer for low-loss and thermally stable graded index plastic optical

fibers. *J. Polym. Sci., Part A: Polym. Chem.*, **47** (13), 3352–3361.
18. Lou, L., Koike, Y., and Okamoto, Y. (2012) A novel copolymer of methyl methacrylate with N-pentafluorophenyl maleimide: high glass transition temperature and highly transparent polymer. *Polymer*, **52** (16), 3560–3564.
19. Kaino, T., Fujiki, M., and Nara, S. (1981) Low-loss polystyrene core-optical fibers. *J. Appl. Phys.*, **52** (12), 7061–7064.
20. Akimoto, Y., Asai, M., Koike, K., Makino, K., and Koike, Y. (2012) Poly(styrene)-based graded-index plastic optical fiber for home networks. *Opt. Lett.*, **37** (11), 1853–1855.
21. Lou, L., Koike, Y., and Okamoto, Y. (2010) Synthesis and properties of copolymers of methyl methacrylate with 2,3,4,5,6-pentafluoro and 4-trifluoromethyl 2,3,5,6-tetrafluoro styrenes: an intrachain interaction between methyl ester and fluoro aromatic. *J. Polym. Sci., Part A: Polym. Chem.*, **48** (22), 4938–4942.
22. Teng, H., Lou, L., Koike, K., Koike, Y., and Okamoto, Y. (2011) Synthesis and characterization of trifluoromethyl substituted styrene polymers and copolymers with methacrylates: effects of trifluoromethyl substituent on styrene. *Polymer*, **52** (4), 949–953.
23. Koike, K., Teng, H., Koike, Y., and Okamoto, Y. (2011) Trifluoromethyl-substituted styrene-based polymer optical fibers. Proceeding of the 20th International Conference on Plastic Optical Fibers, Bilbao, Spain, September 14–16, 2011, p. 031.
24. Resnick, P.R. and Buck, W.H. (1997) in *Modern Fluoropolymers: High Performance Polymers for Diverse Applications* (ed J. Scheirs), John Wiley & Sons, Ltd, Chichester, pp. 397–419.
25. Nakamura, M., Sugiyama, N., Etoh, Y., Aosaki, K., and Endo, J. (2001) Development of perfluoro transparent resins obtained by radical cyclopolymerization for leading edge electronic and optical applications. *Nippon Kagaku Kaishi*, **12**, 659–668.
26. Liu, W., Mikes, F., Guo, Y., Koike, Y., and Okamoto, Y. (2004) Free-radical polymerization of dioxolane and dioxane derivatives: effect of fluorine substituents on the ring opening polymerization. *J. Polym. Sci., Part A: Polym. Chem.*, **42** (20), 5180–5188.
27. Mikes, F., Yang, Y., Teraoka, I., Ishigure, T., Koike, Y., and Okamoto, Y. (2005) Synthesis and characterization of an amorphous perfluoropolymer: Poly(perfluoro-2-methylene-4-methyl-1,3-dioxolane). *Macromolecules*, **38** (10), 4237–4245.
28. Liu, W., Koike, Y., and Okamoto, Y. (2005) Synthesis and radical polymerization of perfluoro-2-methylene-1,3-dioxolanes. *Macromolecules*, **38** (23), 9466–9473.
29. Yang, Y., Mikes, F., Yang, L., Liu, W., Koike, Y., and Okamoto, Y. (2006) Investigation of homopolymerization rate of perfluoro-4,5-substituted-2-methylene-1,3-dioxolane derivatives and properties of the polymers. *J. Fluorine Chem.*, **127** (2), 277–281.
30. Okamoto, Y., Mikes, F., and Koike, Y. (2007) The effect of fluorine substituents on the polymerization mechanism of 2-methylene-1,3-dioxolane and properties of the polymer products. *J. Fluorine Chem.*, **128** (3), 202–206.
31. Muratoni, E., Saito, S., Sawaguchi, M., Yamamoto, H., Nakajima, Y., Miyajima, T., and Okazoe, T. (2007) Synthesis and polymerization of a novel perfluorinated monomer. *J. Fluorine Chem.*, **128** (10), 1131–1136.
32. Sawada, H. (1996) Fluorinated peroxide. *Chem. Rev.*, **96** (5), 1779–1808.
33. Mikes, F., Baldrian, J., Teng, H., Koike, Y., and Okamoto, Y. (2011) Characterization and properties of semicrystalline and amorphous perfluoropolymer: poly(perfluoro-2-methylene-1,3-dioxolane). *Polym. Adv. Technol.*, **22** (8), 1272–1277.
34. Zhang, B., Li, L., Mikes, F., Koike, Y., Okamoto, Y., and Rinaldi, P.L. (2013) Multidimensional NMR characterization of perfluorinated monomer and its precursors. *J. Fluorine Chem.*, **147**, 40–48.
35. Yang, Y., Mikes, F., Yang, L., Liu, W., Koike, Y., and Okamoto, Y. (2006) Novel amorphous perfluorocopolymeric

system: copolymers of perfluoro-2-methylene-1,3-dioxolane derivatives. *J. Polym. Sci., Part A: Polym. Chem.*, **44** (5), 1613–1618.

36. Mikes, F., Teng, H., Kostov, G., Ameduri, B., Koike, Y., and Okamoto, Y. (2009) Synthesis and characterization of perfluoro-3-methylene-2,4-dioxabicyclo[3,3,0] octane: homo- and copolymerization with fluorovinyl. *J. Polym. Sci., Part A: Polym. Chem.*, **47** (23), 6571–6578.

5
Fabrication Techniques

Fabrication processes of POFs (plastic optical fibers), regardless of their step-index (SI) or graded-index (GI) profiles, are divided into two main categories: a preform-drawing process and an extrusion process. The extrusion process is further classified into batch extrusion and continuous extrusion. Currently, most SI POF suppliers employ the continuous extrusion process. Section 5.1 summarizes how POFs can be prepared by each process. In contrast to SI POFs, which have been prepared by employing extruders during the early stages of their development, studies of GI POFs started by exploring how to form a desired graded refractive index profile in a cylindrical polymer rod. Section 5.2 discusses several representative techniques for fabricating GI preforms. The preform-drawing method enables a precise control of the fabrication process of rather complicated refractive index profiles. However, these techniques have not been widely accepted in industry because of their batch nature. More recently, GI POFs have also become producible by an extrusion method. In particular, a dopant diffusion method, which can be adopted for a variety of base polymer materials, is employed by a couple of POF companies in Japan and the United States. Finally, Section 5.3 describes the mechanism and a recently developed approach to simulate the refractive index profile obtained by the dopant-diffusion extrusion process.

5.1
Production Processes of POFs

5.1.1
Preform Drawing

The preform-drawing method is a batch process where a polymeric preform is fabricated first, which is then followed by thermal drawing of the preform into the fiber. A schematic diagram of the process is shown in Figure 5.1. In this method, a cylindrical polymer rod consisting of a core and cladding layers, usually prepared by radical polymerization in bulk under a clean environment, is positioned vertically in the middle of the furnace where its lower portion is heated locally to the drawing temperature. The furnace temperature appropriate for the preform is

Fundamentals of Plastic Optical Fibers, First Edition. Yasuhiro Koike.

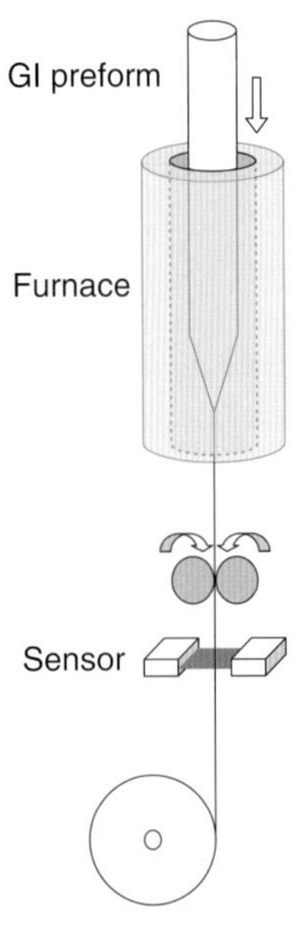

Figure 5.1 Preform-drawing process.

dependent on the viscoelasticity of the base polymer, which is also in turn highly dependent on the molecular weight and the concentration of any additive used. In addition, it depends on how fast the fiber needs to be produced. The preform tip softens in the furnace and gradually deforms and falls under its own weight. The tip is cut and the drawn fiber is wound onto a take-up reel. In the meantime, the preform is slowly lowered through the furnace. The diameter of the fiber, which is continuously monitored by a sensor, can be controlled by the feeding speed of the preform and the winding speed of the fiber. Though the diameter of the fiber fluctuates for a while at the beginning, it gradually stabilizes to the desired size.

The fabrication methods of SI preforms are not discussed in this book. Literature on these methods is widely available but not in strong demand today, both in academia and industry, except for some specific applications. Several representative techniques for fabricating GI preforms are discussed in Section 5.2.

5.1.2
Batch Extrusion

The batch extrusion process, which produces POFs in a completely closed system from the distillation of a monomer by fiber drawing, was developed by Kaino *et al.* [1]. Figure 5.2 shows a schematic diagram of the apparatus. A purified monomer is placed in a monomer flask, and an initiator and a chain-transfer agent are placed in another flask. Both flasks are sealed and evacuated. After freeze–pump–thaw cycles, the monomer and the mixture of the initiator and chain-transfer agent are distilled into the polymerization reactor. The reactor is heated at a certain temperature for a period adequate to fully polymerize the

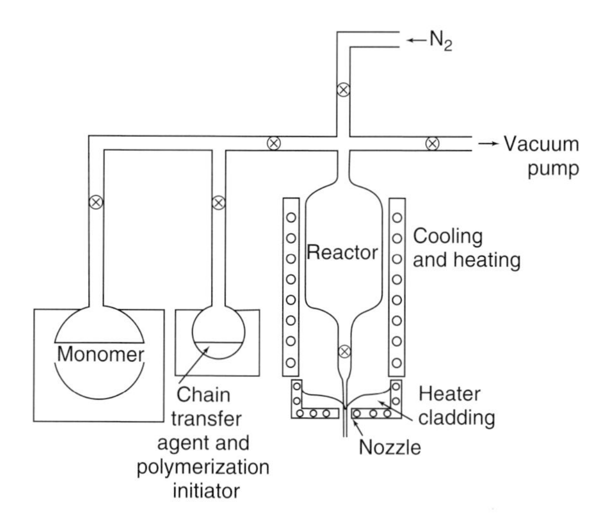

Figure 5.2 Batch extrusion process.

monomer. Once the polymerization is complete, the reactor is further heated until the polymer becomes a fluid. Dry nitrogen gas is used to pressurize the reactor, which pushes the material through a nozzle to form the fiber. The fiber is pulled through another nozzle where it is coated with a cladding polymer. It was in this way that low-loss SI POFs were successfully fabricated. In 1981, Kaino *et al.* [1] reported a polystyrene (PSt)-based SI POF with an attenuation value of 114 dB/km at 670 nm, and later succeeded in obtaining a poly(methyl methacrylate) (PMMA)-based SI POF with an attenuation of 55 dB/km at 568 nm [2].

5.1.3
Continuous Extrusion

Today, most SI POFs are produced by a continuous extrusion process, which greatly improves the productivity of batch extrusion. Figure 5.3 shows a schematic diagram of this process, first developed by Mitsubishi Rayon in 1974 [3]. In this method, a purified monomer, along with a free-radical initiator and a chain-transfer agent, is fed into a reaction chamber where the polymerization reaction takes place. The polymerization conversion is controlled at around 60–80%. Because of the presence of the unreacted monomer, the polymer solution can be transported at a considerably lower temperature than that of the pure polymer. The polymer syrup, which will be the core of the SI POF, is then fed into a devolatilizing extruder using a gear pump. The remaining monomer is evaporated and returned to the polymerization reactor for further use. Subsequently, the core polymer, whose conversion has reached almost 100%, runs into a coextrusion die where a cladding polymer fed by a separate extruder is coated onto the core polymer. Finally, the extruded polymer with the core and cladding is drawn into a fiber.

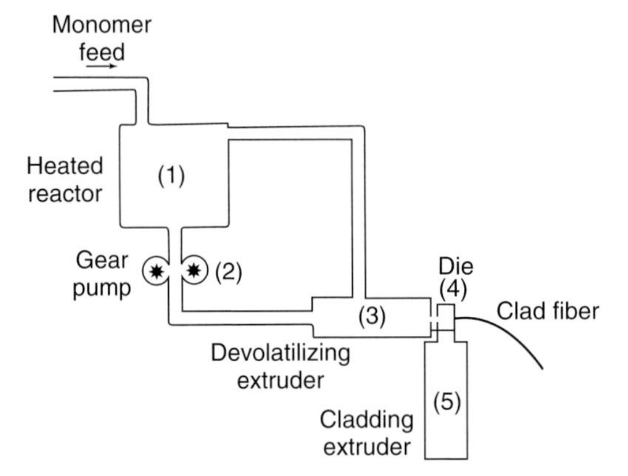

Figure 5.3 Continuous extrusion process.

5.2 Fabrication Techniques of Graded-Index Preforms

5.2.1 Copolymerization

The initial methods to fabricate GI preforms took advantage of the difference in monomer reactivity ratios in copolymerization reactions. These methods are of two types: photo-copolymerization [4–6] and interfacial-gel copolymerization [7, 8].

In copolymerization reactions, the monomer reactivity ratio is often referred to in order to discuss how the reaction proceeds. The monomer reactivity ratios r_{12} and r_{21} are defined as the ratios of the rate constant of the reaction of a given radical with its own monomer (M_1) to the rate constant of its addition to the other monomer (M_2). Thus, $r_{12} > 1$ means that the radical $M_1^{\cdot}$ prefers the addition of M_1, whereas $r_{12} < 1$ means that it prefers the addition of M_2. In these copolymerization methods, comonomers satisfying the following requirements are used as the base material: $r_{12} > 1$, $r_{21} < 1$, and the refractive index of the M_1 homopolymer is lower than that of the M_2 homopolymer. More details about monomer reactivity ratios are described later in this chapter.

A schematic diagram of the photo-copolymerization process is shown in Figure 5.4. The monomer mixture, containing a specified amount of a photoinitiator and a thermal initiator, is sealed inside a glass tube. The tube is mounted on a rotating table that turns the tube around its axis. A UV lamp equipped with shades is mounted on a vertical translation stage that moves upward at a constant velocity. Here, the vertical speed of the lamp is adjusted to ensure that the entire monomer in the tube is prepolymerized after the light source traverses its length. The reaction is carried out at the bottom to avoid the formation of a cavity by volume shrinkage during copolymerization. Once the comonomer is exposed to UV light, the copolymerization reaction starts from the outside where the light

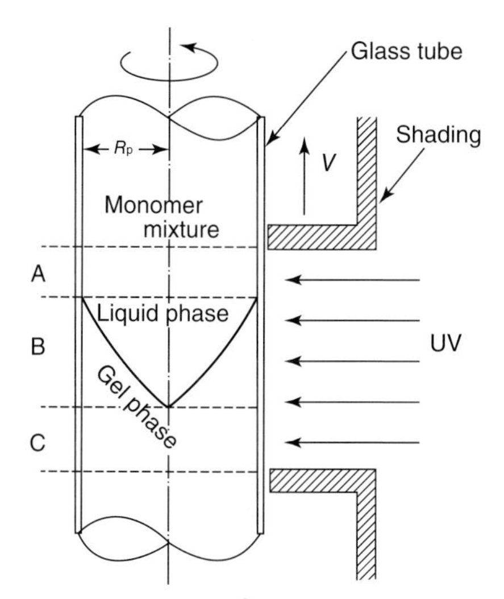

Figure 5.4 Photo-copolymerization process.

intensity is the strongest, and a thin gel phase is formed along the inner wall; the reaction occurs preferentially in the gel phase. Consequently, the copolymer phase moves inward toward the center as the reaction proceeds. Here, because M_1 is more reactive than M_2, the proportion of M_1 at the periphery is greater than that in the middle. As a result, the refractive index of the rod decreases with distance from the center in accordance with the change in the copolymer composition (M_1 has a lower refractive index). After the photo-copolymerization process, the rod is heat-treated to fully polymerize the remaining monomer. The GI preform is removed from the glass tube and heat-drawn into a fiber.

GI preforms can also be fabricated by an interfacial-gel polymerization technique. The principle is basically the same as that of the photo-copolymerization method discussed above, except for the mechanism that forms the initial gel phase. In this method, the core solution (the monomer) is placed in a polymer tube rather than in a glass tube. The gel phase in the photo-copolymerization method is referred to as a *prepolymer* with a conversion of less than 100%, whereas in this method the gel phase comprises the polymer layer on the inner wall of the tube swollen by the core monomer. The reaction is carried out under UV irradiation or heating.

In copolymerization reactions between M_i and M_j monomers, the monomer reactivity ratio r_{ij} is defined as follows:

$$r_{ij} = \frac{k_{ii}}{k_{ij}} \qquad \begin{matrix} i = 1, 2, \ldots, n \\ j = 1, 2, \ldots, n \end{matrix} \qquad i \neq j, \tag{5.1}$$

where k_{ii} and k_{ij} are the propagation rate constants in the following copolymerization reactions:

$$M_i^{\cdot} + M_i \xrightarrow{k_{ii}} M_i M_i^{\cdot}$$

$$M_i^{\cdot} + M_j \xrightarrow{k_{ij}} M_i M_i^{\cdot}$$

The monomer reactivity ratios r_{ij} and r_{ji} between the M_i and M_j monomers are estimated by

$$r_{ij} = \left(\frac{Q_i}{Q_j}\right) \exp[-e_i(e_i - e_j)]$$
$$r_{ji} = \left(\frac{Q_j}{Q_i}\right) \exp[-e_j(e_j - e_i)]. \tag{5.2}$$

Here, Q_i (or Q_j) is the reactivity of the monomer M_i (or M_j), and e_i (or e_j) is the electrostatic interaction of the permanent charges on the substituents that polarize the vinyl group of M_i (or M_j). Table 5.1 shows the refractive indices of representative polymers for radical polymerizations as well as the Q and e values of the monomers. These monomers are categorized into nine groups as follows: MA, methacrylate; XMA, methacrylate with high refractive index; A, acrylate; XA, acrylate with high refractive index; Ac, vinyl acetate; XAc, vinyl acetate with high refractive index; C, acrylonitrile; α-A, α-substituted acrylate; St, styrene. The refractive index profile of GI POF is formed by copolymerizing two or three specific types of monomers in Table 5.1 based on the following concepts.

5.2.1.1 Binary Monomer System

In order to obtain GI profiles by copolymerization, two monomers (M_1 and M_2) satisfying the following conditions need to be employed:

$$r_{12} > 1 \quad \text{and} \quad r_{21} < 1$$
$$n_1 < n_2 \tag{5.3}$$

where n_1 and n_2 are the refractive indices of the M_1 and M_2 homopolymers, respectively. Because the reaction proceeds from the periphery to the center in these copolymerization methods, the outer region of the obtained preform contains more M_1 unit whereas the middle part contains more M_2 unit. Thus, the refractive index of the preform decreases from the center to the periphery.

Figure 5.5 shows a Q–e map of the monomers listed in Table 5.1. When MMA (methyl methacrylate) is used as the M_1 monomer, the regions of ($r_{12} < 1$ and $r_{21} > 1$), ($r_{12} < 1$ and $r_{21} < 1$), and ($r_{12} > 1$ and $r_{21} < 1$) are separated by the solid curves. From the figure, candidates for the M_2 monomer that satisfy the monomer reactivity ratios and the refractive index of Equation 5.3 are narrowed down to XA and XAc groups. Figure 5.6 shows the representative refractive index profiles of GI preforms based on MMA–benzyl acrylate (BzA), MMA–vinyl benzoate (VB), and MMA–vinyl phenyl acetate (VPAc) copolymers.

Table 5.1 Categories of possible monomers for photo-copolymerization and interfacial-gel copolymerization processes.

Group	Monomer			n_D of polymer	Q	e
MA	1	MMA	Methyl methacrylate	1.49	0.74	0.40
	2	EMA	Ethyl methacrylate	1.483	0.73	0.52
	3	*n*PMA	*n*-Propyl methacrylate	1.484	0.65	0.44
	4	*n*BMA	*n*-Butyl methacrylate	1.483	0.78	0.51
	5	*n*HMA	*n*-Hexyl methacrylate	1.481	0.67	0.34
	6	*i*PMA	Isopropyl methacrylate	1.473	0.85	0.34
	7	*i*BMA	Isobutyl methacrylate	1.477	0.72	0.24
	8	*t*BMA	*t*-Butyl methacrylate	1.463	0.76	0.24
	9	CHMA	Cyclohexyl methacrylate	1.507	0.82	0.45
XMA	10	BzMA	Benzyl methacrylate	1.568	0.75	0.25
	11	PhMA	Phenyl methacrylate	1.57	1.49	0.73
	12	1PhEMA	1-Phenylethyl methacrylate	1.549	0.74	0.36
	13	2PhEMA	2-Phenylethyl methacrylate	1.559	0.74	0.36
	14	FFMA	Furfuryl methacrylate	1.538	0.78	0.04
A	15	MA	Methyl acrylate	1.4725	0.42	0.60
	16	EA	Ethyl acrylate	1.4685	0.52	0.22
	17	BA	*n*-Butyl acrylate	1.4634	0.50	1.06
XA	18	BzA	Benzyl acrylate	1.5584	0.34	0.90
	19	2ClEA	2-Chloroethyl acrylate	1.52	0.41	0.54
Ac	20	VAc	Vinyl acetate	1.47	0.026	−0.22
XAc	21	VB	Vinyl benzoate	1.578	0.061	−0.55
	22	VPAc	Vinyl phenylacetate	1.567	0.018	−1.078
	23	VClAc	Vinyl chloroacetate	1.512	0.074	−0.65
C	24	AN	Acrylonitrile	1.52	0.60	1.20
	25	αMAN	α-Methylacrylonitrile	1.52	1.12	0.81
α-A	26	MA(2Cl)	Methyl-*a*-chloroacrylate	1.5172	2.02	0.77
	—	MAtro	Atropic acid, methyl ester	1.56	4.78	1.20
St	27	*o*-ClSt	*o*-Chlorostyrene	1.6098	1.28	−0.36
	28	*p*-FSt	*p*-Fluorostyrene	1.566	0.83	−0.12
	29	o, *p*-FSt	*o*, *p*-Difluorostyrene	1.475	0.65	−0.31
	—	*p*-*i*PSt	*p*-Isopropyl styrene	1.554	1.60	−1.52

5.2.1.2 Ternary Monomer System

When using three different types of monomers, not only the GI profiles but also various other profiles are obtained when the monomers satisfy the following conditions:

$$\begin{aligned} &r_{12} > 1 \quad r_{13} > 1 \quad r_{23} > 1 \\ &r_{21} < 1 \quad r_{31} < 1 \quad r_{32} < 1 \end{aligned} \tag{5.4}$$

An M_1-rich terpolymer is formed in the initial stage of copolymerization, an M_2-rich terpolymer in the intermediate stage, and an M_3-rich terpolymer in the final stage. Because the terpolymer phase grows inward toward the center axis, the composition changes from M_1-richer terpolymer to M_3-richer

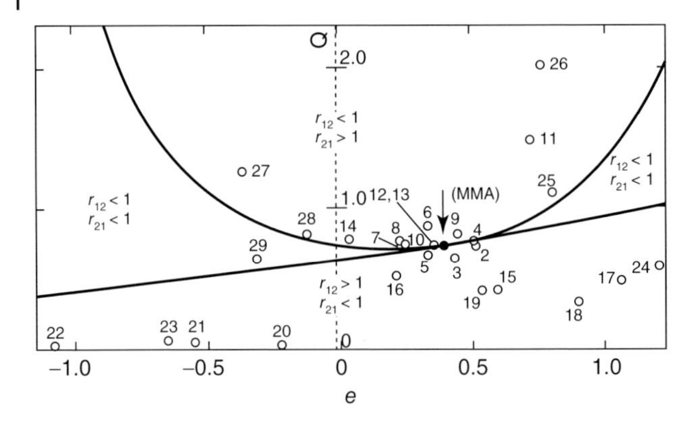

Figure 5.5 *Q*–*e* map of monomers shown in Table 5.1: 1, MMA; 2, EMA; 3, *n*PMA; 4, *n*BMA; 5, *n*HMA; 6, *i*PMA; 7, *i*BMA; 8, *t*BMA; 9, CHMA; 10, BzMA; 11, PhMA; 12, 1PhEMA; 13, 2PhEMA; 14, FFMA; 15, MA; 16, EA; 17, BA; 18, BzA; 19, 2ClEA; 20, VAc; 21, VB; 22, VPAc; 23, VClAc; 24, AN; 25, αMAN; 26, MA(2Cl); 27, *o*-ClSt; 28, *p*-FSt; and 29, *o*, *p*-FSt.

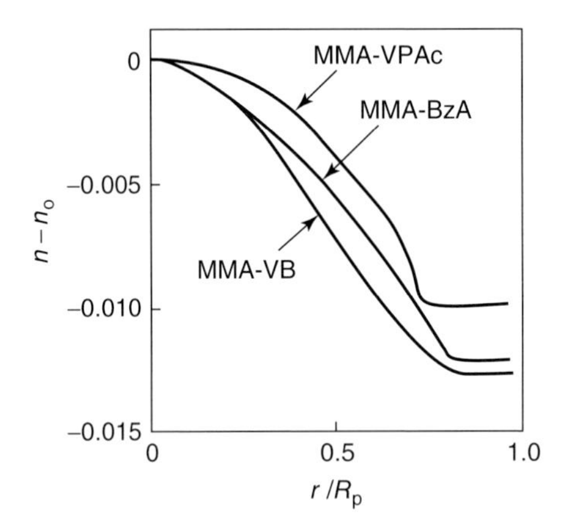

Figure 5.6 Refractive index profiles of GI preforms prepared by the interfacial-gel copolymerization process. n_0 and n are the refractive indices at the center and at distance r from the central axis, respectively, and R_p is the radius of the rod. Monomer feeding ratio is $M_1/M_2 = 4$ (wt/wt).

terpolymer as the reaction proceeds. The ternary monomer system is classified into six types based on the refractive index of each homopolymer, as shown in Figure 5.7. Each black dot in the figure indicates the M_1, M_2, and M_3 homopolymers from the left. The ordinate is the refractive index of each homopolymer. When monomer systems IV and V are used, a W-shaped profile is obtained. Figure 5.8 shows a representative refractive index profile of a W-shaped preform based on the BzMA–VAc–VPAc ternary monomer system prepared by the photo-copolymerization method.

These profiles can be predicted prior to the actual fabrications. In the multiple monomer system (M_1, M_2, … , M_n), the differential equation of the copolymer

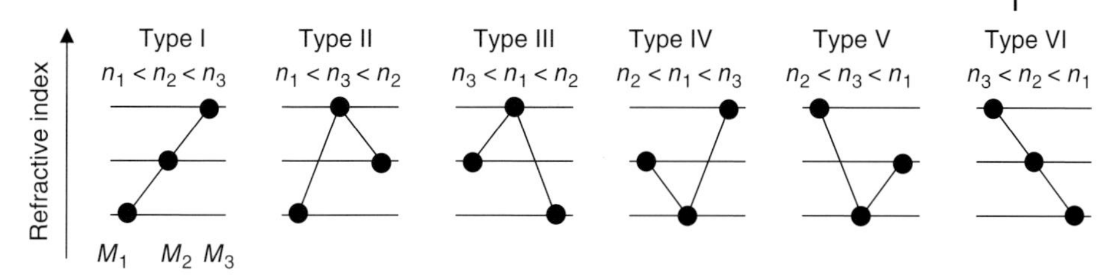

Figure 5.7 Classification of ternary monomer systems.

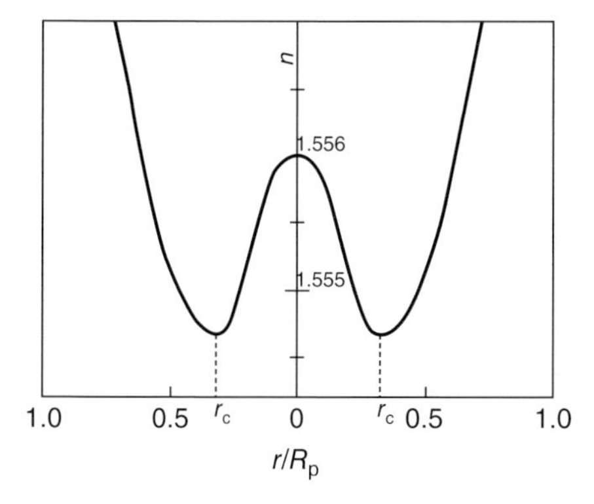

Figure 5.8 Refractive index distribution of a W-shaped GI preform rod prepared by photo-copolymerization of the BzMA–VAc–VPAc monomer system.

composition is expressed as follows:

$$\frac{d[M_i]}{d[M_j]} = \frac{\left(D_{ii}\sum_{k=1}^{n}\frac{[M_k]}{r_{ik}}\right)}{\left(D_{jj}\sum_{k=1}^{n}\frac{[M_k]}{r_{jk}}\right)} \tag{5.5}$$

Here, $d[M_i]/d[M_j]$ is the molar ratio of the M_i monomer unit to the M_j monomer unit in the copolymer formed at any instant when the concentration of each monomer (M_1, M_2, … , M_n) in the monomer mixture is $[M_1]$, $[M_2]$, … , $[M_n]$, respectively. The value r_{ik} (r_{jk}) is the monomer reactivity ratio between the M_i (M_j) and M_k monomers, and $r_{ii} = r_{jj} = 1$. D_{ii} (D_{jj}) is a determinant in which the i line and i column (j line and j column) have been omitted from the determinant D:

$$D = \begin{vmatrix} [M_1] - \sum_{k=1}^{n}\frac{[M_k]}{r_{1k}} & \cdots & \frac{[M_1]}{r_{n1}} \\ \vdots & \ddots & \vdots \\ \frac{[M_n]}{r_{1n}} & \cdots & [M_n] - \sum_{k=1}^{n}\frac{[M_k]}{r_{nk}} \end{vmatrix}. \tag{5.6}$$

The weight fraction (y_k) of the M_k monomer unit in the copolymer is expressed as

$$y_k = m_k \frac{d[M_k]}{\sum_{k=1}^{n} m_k d[M_k]}, \tag{5.7}$$

where m_k is the molecular weight of the M_k monomer. Because the number of monomer reactivity ratios is $n(n-1)$, ternary monomer systems have more parameters than binary systems.

By solving the differential Equation (5.5) and relating it to the conversion (P) from the monomer to the polymer, the change in the copolymer composition during the reaction of a multiple monomer system can be calculated. The weight fraction (x_k) of the M_k monomer ($k = 1, 2, \ldots, n$) in the remaining monomer mixture with conversion P is expressed as

$$x_k = \frac{x_{k0} - \int_0^P y_k dP}{1 - P}, \tag{5.8}$$

where x_{k0} denotes the weight fraction of M_k in the initial monomer mixture. From Equations 5.5–5.8, the copolymer composition of the instantaneous copolymer formed with conversion P is estimated. Furthermore, using the Lorentz–Lorenz equation and assuming the additivity of the molar volumes of structural units, the refractive index n of the copolymer, composed of $d[M_1]$, $d[M_2]$, … , $d[M_n]$, is expressed as

$$n = \sqrt{\frac{1 + 2\varphi}{1 - \varphi}}, \tag{5.9}$$

$$\varphi = \frac{\sum_{k=1}^{n} \left(\frac{{n_k}^2 - 1}{{n_k}^2 + 2} \cdot \frac{m_k d\left[M_k\right]}{\rho_{pk}} \right)}{\sum_{k=1}^{n} \left(\frac{m_k d\left[M_k\right]}{\rho_{pk}} \right)},$$

where n_k and ρ_{pk} are the refractive index and density of the M_k homopolymer, respectively.

The variation of the calculated terpolymer composition with conversion P in the MMA–BzA–VB system is shown in Figure 5.9, where $r_{12} = 2.66$, $r_{21} = 0.29$, $r_{13} = 8.52$, $r_{31} = 0.07$, $r_{23} = 1.51$, and $r_{32} = 0.08$. Each plot expresses the instantaneous copolymer composition at $P = 5k$ wt% ($k = 1, 2, \ldots$). This trend indicates that the MMA-rich terpolymer is formed in the initial stage, the BzA-rich terpolymer in the intermediate stage, and the VB-rich terpolymer in the final stage. The calculated refractive index in this terpolymer system is shown in Figure 5.10. The refractive indices of the MMA, BzA, and VB homopolymers are 1.49, 1.56, and 1.58, respectively. In the binary monomer system (curve A), the refractive index abruptly increases when $P \approx 80$ wt%. Addition of BzA (ternary system) causes a gentle increase in the refractive index, as shown in curves B, C, and D.

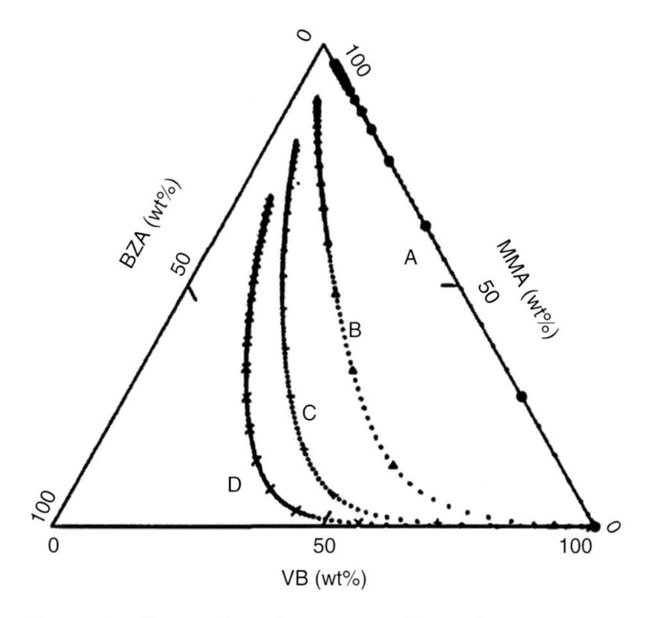

Figure 5.9 Change of copolymer composition with conversion, P, in the MMA–BzA–VB system. Each plot represents instantaneous copolymer composition at $P = 5k\,\text{wt\%}(k = 1, 2, \ldots)$. MMA–BzA–VB (wt/wt/wt): A, 3/0/1; B, 2.5/0.5/1; C, 2/1/1; and D, 1.5/1.5/1.

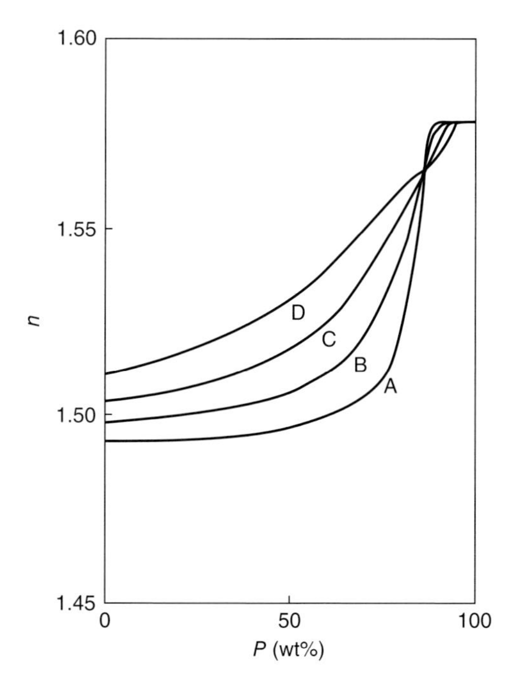

Figure 5.10 Refractive index of the MMA–BzA–VB copolymer as a function of conversion P. MMA–BzA–VB (wt/wt/wt): A, 3/0/1; B, 2.5/0.5/1; C, 2/1/1; and D, 1.5/1.5/1.

5.2.2
Preferential Dopant Diffusion

The second method to obtain GI profiles in a cylindrical rod uses a high-refractive-index dopant instead of the copolymerization reaction. GI POFs prepared by the previous method tend to have high attenuation because the fluctuation of the dielectric constant caused by the copolymerization of monomers with different reactivities can cause considerable light scattering loss. As discussed in Section 2.2.2, light scattering due to large heterogeneous structures V_{V2}^{iso} shows an angular dependence and is defined as follows [9]:

$$V_{V2}^{iso} = \frac{8\pi^3\eta^2a^3}{\lambda_0^4(1+k^2s^2D^2)^2}, \tag{5.10}$$

$$k = \frac{2\pi}{\lambda},$$

$$s = 2\sin\left(\frac{\theta}{2}\right).$$

Here, η^2 is the mean-square average of the fluctuations of all dielectric constants, a is the correlation length, λ and λ_0 are the wavelengths of light in a specimen and under vacuum, respectively, and θ is the scattering angle with respect to the direction of the incident ray. This equation indicates that V_{V2}^{iso} increases as the correlation length a increases. When a two-phase model as described in Figure 4.5 is assumed, the correlation length D, which is a measure of the amount of heterogeneity, is given by [10]

$$D = \frac{4V}{S}V_AV_B, \tag{5.11}$$

where S is the total contact area of the A and B phases, which have different refractive indices, V is the total volume, and V_A and V_B are the volume fractions of the A and B phases ($V_A + V_B + 1$), respectively. When V_A and V_B are constant, the correlation length a decreases with increasing contact area between the A and B phases, as described in Figure 4.5b. Consequently, by decreasing the volume of each component of the A phase, light scattering is inhibited.

In this method, nonreactive compounds are employed as the high-refractive-index component [11]. For example, MMA and bromobenzene (BB), which have higher refractive indices than PMMA, can be utilized as the monomer and the nonreactive compound, respectively. The fabrication procedure is the same as in the photo-copolymerization and interfacial-gel polymerization methods. However, the principle of formation the GI profile is different. In contrast to the previous methods that use the difference in the monomer reactivity ratios, in this method the difference in the molecular size is important. Because the molecular size of MMA is smaller than that of BB, MMA more easily diffuses into the gel phase. Thus, BB molecules are concentrated into the middle region to form the GI profile as the polymerization progresses. The mechanism is schematically described in Figure 5.11.

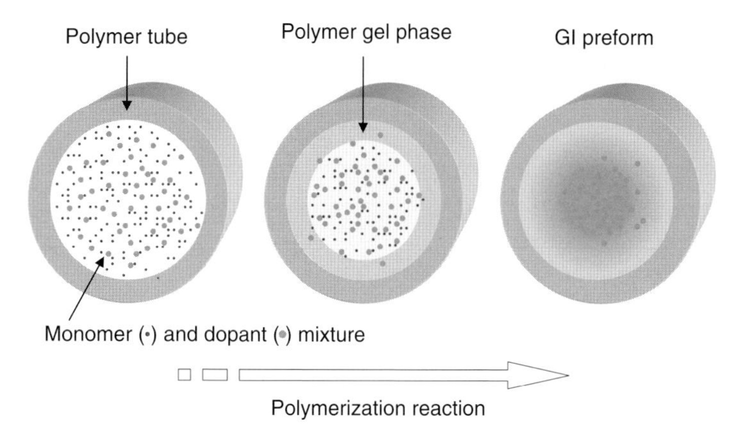

Figure 5.11 Interfacial-gel polymerization technique using dopants.

Subsequent to this achievement, numerous dopants have been explored to improve the properties of GI POFs. Ideal dopants are relatively low-molecular-weight compounds that (i) are soluble in base polymers and do not phase-separate or crystallize over time, (ii) do not significantly increase the attenuation of the polymers, (iii) do not reduce the glass transition temperature (T_g) of the polymers by an unacceptable degree, (iv) provide large changes in the refractive indices at low concentrations, for example, $n_1 - n_2 > 0.015$ for less than 15 wt% dopant, (v) are chemically stable in the polymers at the processing temperatures, (vi) have low volatility at the processing temperatures, and (vii) are substantially immobilized in the glassy polymer in the operating environment. In particular, compared to previous copolymer systems, the polymer–dopant system has the big concern that the dopant significantly decreases T_g because of the plasticization effect. Here, T_g of a noncrystalline material is the critical temperature at which the material changes its behavior from being glassy to rubbery, and vice versa. Because the operating temperature of a POF is basically dependent on its T_g, this is considered to be one of the most important physical properties. Thus, it is preferable that the refractive index of the dopant be as high as possible so that a lower dopant concentration could form a GI profile with a sufficiently large numerical aperture (NA). For PMMA-based GI POFs, diphenyl sulfide (DPS) has proven to be the most suitable dopant identified so far. By using dopants for the formation of GI profiles, the attenuation of GI POFs was drastically reduced to the same level as that of SI POFs.

5.2.3 Thermal Dopant Diffusion

The latest GI-preform method is the rod-in-tube method [12]. The idea, shown in Figure 5.12, was originally developed from a fabrication method of glass optical

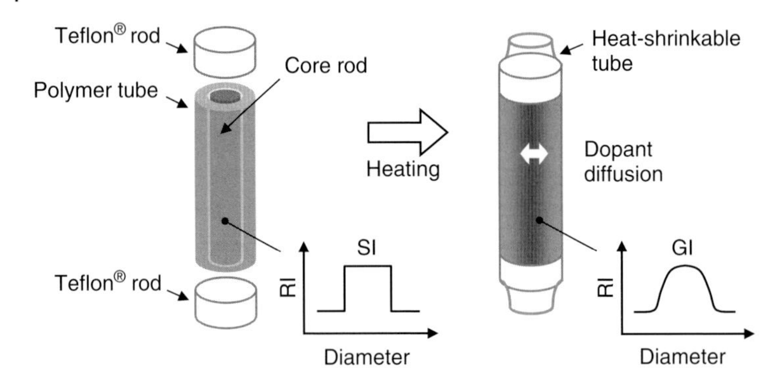

Figure 5.12 Rod-in-tube method.

fibers. In this method, the core rod containing a high-refractive-index dopant and the cladding tube are prepared separately. The outer diameter of the rod and the inner diameter of the tube are designed to be as close as possible, whereas the height of the rod is slightly more than that of the tube so that any water and air in the interface can be easily removed. After washing with purified water, the rod is inserted into the tube, which is then covered with a heat-shrinkable tube. The original diameter of the heat-shrinkable tube is ~1.2 times the outer diameter of the cladding tube. More importantly, when shrunk, the size of the heat-shrinkable tube must be smaller than the combined value of the core-rod diameter and the thickness of the cladding tube. A Teflon® disk having almost the same diameter as the cladding tube is placed at the bottom, and the end is completely closed by heating. Another Teflon disk is then placed on the other end, and the assembly is placed vertically in a vacuum oven. Note that both the core rod and the cladding tube are still exposed to the air at this moment. After drying under vacuum at 40–50 °C for 2 h, the assembly is heated to a temperature above the softening points of each polymer. During the heat treatment, the core rod and cladding tube adhere to each other because of the contractile force of the heat-shrinkable tube, and the dopant gradually diffuses into the cladding layer, forming the GI profile.

The greatest advantage of this method is that GI preforms can be obtained from any core and cladding polymer combination as long as the two adhere well. In the case of the interfacial-gel polymerization technique, the core monomer is polymerized in the presence of a dissolved cladding polymer. Consequently, the core–cladding boundary becomes ambiguous unless the polymers are compatible and the refractive index difference is negligibly small. However, in the rod-in-tube method, the core and cladding layers are physically attached to each other, and the GI profile is formed by distributing the dopant in the radial direction. Thus, there is no concern about an increase in light scattering. At present, this method is most often utilized for laboratory-scale experiments.

5.2.4
Polymerization under Centrifugal Force

The last method used to prepare GI preforms employs centrifugal force. This method is divided into two categories based on the role of centrifugal force in forming the GI profiles.

The first category of the method was developed by Van Duijnhoven *et al.* in 1999 [13, 14]. In their study, distilled MMA and 2,2,3,3-tetrafluoropropylmethacrylate (TFPMA) were utilized as the base material. First, TFPMA was thermally polymerized with benzoyl peroxide (BPO) in bulk in a vacuum-sealed glass tube. The obtained bulk polymer was dissolved in MMA, and the polymer solution was placed in a glass tube. The polymerization was carried out in the tube rotating at up to 20 000 rpm in an oven at 60–80 °C for 12 h, and the preform was further heated at 120 °C for several hours under continuous rotation. In such a strong centrifugal force field, poly(TFPMA), which is denser than MMA, was driven to the outer periphery. Because the refractive index of poly(TFPMA) is lower than that of PMMA, a GI profile based on the concentration gradient of poly(TFPMA) was formed. Figure 5.13 shows the (a) compositional gradients of poly(TFPMA) in the radial direction of the obtained preforms and (b) refractive index profiles calculated from the compositional gradients. The ratio of these two monomer units in the blend polymer was determined by Raman spectroscopy. The refractive index was calculated from the composition using the

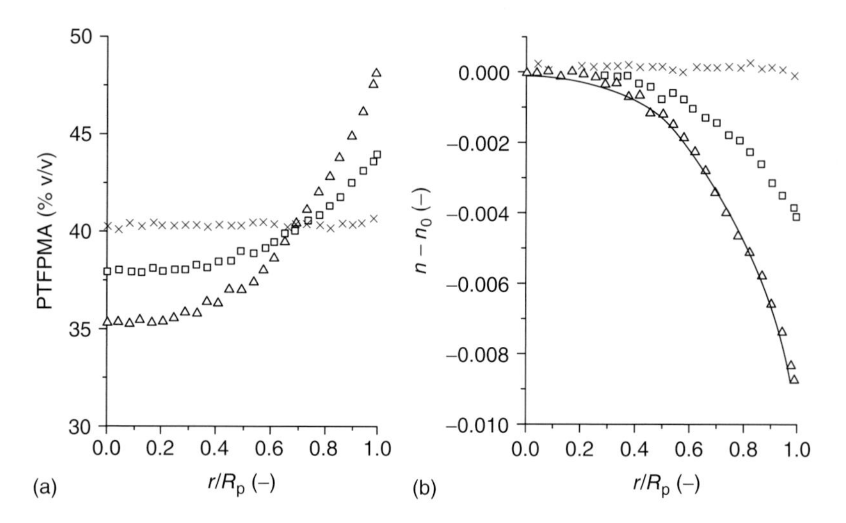

Figure 5.13 Compositional and refractive index gradients along the radial direction of preforms. (a) Volume percent of poly(TFPMA) (expressed as PTFPMA in the figure) and (b) refractive index difference from the center; r is the position in radial direction; R_p is the radius of the preform; n is the refractive index; and n_0 is the refractive index at core center. Preforms were prepared in glass tubes rotating at (×) 0 rpm, (□) 15 000 rpm, and (Δ) 20 000 rpm. Solid line in (b) represents refractive index profile measured obtained by interference microscopy.

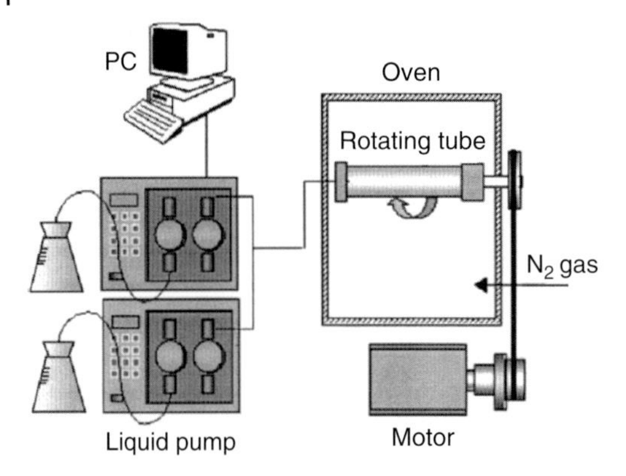

Figure 5.14 Centrifugal deposition method.

Lorentz–Lorenz equation. These results show that poly(TFPMA) and PMMA are not completely separated by the centrifugal force, and the composition of the blend polymer gradually changes along the radial direction. The magnitude of the gradients was also confirmed to be dependent on the rotational speed.

A disadvantage of this method compared to the others discussed above is that the obtained preform inevitably has a cavity along the rotational axis even though the reactor is filled with the polymer solution. This is due to a volume shrinkage associated with the polymerization reaction. Although the cavity could be collapsed during the heat-drawing process under vacuum when the hole size was small enough, the increase in the fiber attenuation was unavoidable. Im *et al.* [15, 16] proposed a modified fabrication method of GI preforms in 2002. They used MMA and St as the base materials and filled the cavity by feeding the comonomers or the polymer solution repeatedly until the cavity was no longer observed. The feeding ratios were determined from the composition of the inner walls of the tubes to obtain continuous refractive index profiles.

The other method, termed the *centrifugal deposition method*, was developed by Shin *et al.* [17] in 2003. The schematic diagram of the experimental setup is shown in Figure 5.14. The reaction system consists of an oven, a rotating tube and its holding part, two feeding pumps, and a feeding nozzle for the reactants. The oven is filled with nitrogen gas, and the rotation speed of the glass tube is 5400 rpm. MMA and benzyl methacrylate (BzMA), which form an ideal random copolymer, were employed as the base material. In this method, certain amounts of MMA and BzMA are repeatedly injected into the tube, which is heated at 80 °C. The feeding ratio of the comonomers is controlled by a computer and is continuously varied from MMA/BzMA = 100/0 to 85/15 vol%, so that a continuous GI profile is obtained from the periphery to the center of the tube. The preform is further heated at 120 °C for 12 h to increase the conversion and is dried at 110 °C for another 12 h under vacuum. In the previous method, the glass tube was rotated at a high speed to separate the materials according to the difference in the density, whereas the centrifugal force in this method is utilized only to form each layer symmetrically. Any part of the obtained preform is an ideal random copolymer with a narrow composition distribution, which should be the same as the feed composition. Thus, the intensity of light scattering due to the heterogeneous structure is considerably lower in this system as compared to the previous blended polymer systems.

5.3 Extrusion of GI POFs

Compared to the preform-drawing processes, extrusions are more productive. The principle of forming GI profiles by extrusion [18] is almost the same as that of the rod-in-tube method discussed above. The general setup is shown in Figure 5.15. All the tooling components are fabricated from corrosion-resistant materials, for example, Hastelloy, and the surfaces that come into contact with the flowing melts are polished to promote a smooth flow with little material stagnation. All parts are machined to a high precision to ensure a good fit and prevent any leakage. The shaded areas in the figure represent heaters, which are

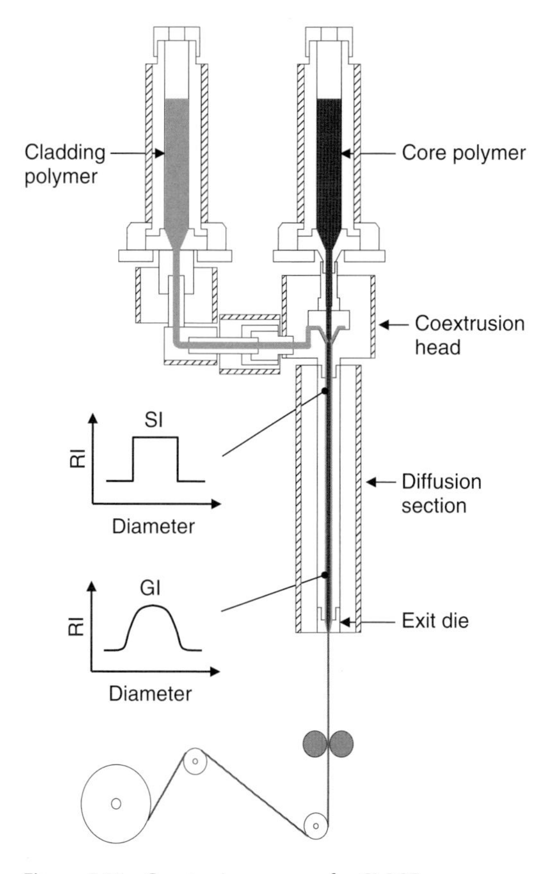

Figure 5.15 Coextrusion process for GI POFs.

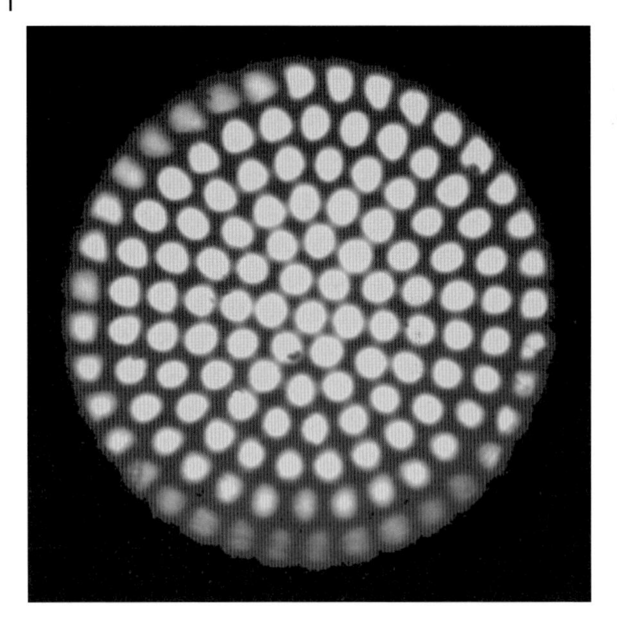

Figure 5.16 Cross-sectional area of a 127-core GI POF. Each core diameter is 25 μm.

individually controlled by a programmable thermostat. The core and cladding polymers are fed into separate extruders, and the core polymer contains one or more diffusible dopants that provide the desired GI. The core polymer is fed from the core extruder through a connector hose into a coextrusion head, and the cladding polymer is fed into the coextrusion head through a connector pipe. The coextrusion head contains a core nozzle for directing the core polymer into the center of the diffusion section, while the cladding polymer from the connector pipe is distributed concentrically around the core polymer into the diffusion section. Here, the number of cores depends on the design of coextrusion head. A multi-core GI POF, which is shown in Figure 5.16, is also easily fabricated using a coextrusion head with multiple core nozzles [19]. The core and cladding polymers flow together into the diffusion zone. The diffusion section is also heated to maintain the flow of the molten polymer and promote diffusion of the dopant; the length of the diffusion zone is designed to allow the desired extent of diffusion to occur. The polymers then flow from the section into an exit die, and the exiting fiber is then pulled by a capstan to draw the fiber at the necessary rate to obtain the desired final diameter. Using the new system, the coextrusion process has become capable of producing GI POFs at commercially useful speeds.

The refractive index profile of GI POFs prepared by the extrusion method can also be predicted prior to the actual fabrication. The diffusion phenomena of dopants in polymers while passing through the diffusion section can be described by the advection–diffusion equation [20]

$$\frac{\partial c}{\partial t} = \frac{1}{r}\frac{\partial c}{\partial r}\left(D_{\mathrm{m}} r \frac{\partial c}{\partial r}\right) - u\frac{\partial c}{\partial z}. \tag{5.12}$$

Here, c is the dopant concentration, t is the diffusion time, D_m is the mutual diffusion coefficient, u is the velocity field, and r and z are the positions in the radial and vertical directions of the diffusion section, respectively. The mutual diffusion coefficient D_m is exponentially dependent on the dopant concentration as

$$D_m(c) = D_0 \exp(\alpha c). \tag{5.13}$$

Here, D_0 is the diffusion coefficient at infinite dilution, and α is the concentration-dependent index. These values can be obtained from a one-dimensional diffusion experiment [21]. In addition, in the case of a Newtonian fluid, the velocity field u is expressed as a function of r by an exponential model as follows:

$$u(r) = 2u_{ave}\left\{1 - \left(\frac{r}{R}\right)^3\right\}. \tag{5.14}$$

Here, u_{ave} is the average velocity corresponding to the discharge rate of the polymer from the diffusion section, and R is the inner radius of the diffusion tube. By substituting Equations 5.13 and 5.14 into Equation 5.12, the refractive index profiles of the GI POFs can be calculated.

As seen from these equations, there are several parameters can be adjusted to obtain a desired profile. Figure 5.17 is one example, showing that the index exponent g can be controlled by changing the length of the diffusion section and the discharge rate of the polymer. The base polymer and the dopant are poly(2,2,2-trifluoroethyl methacrylate) (poly(TFEMA)) and DPS, respectively. The broken line represents the optimum g value that minimizes the summation of the modal and material dispersions. The figure indicates that the optimum refractive index profile should be formed after the polymer passes through a diffusion section of 350 mm at a rate of 0.44 g/min. Figure 5.18 shows the result. The GI POF fabricated

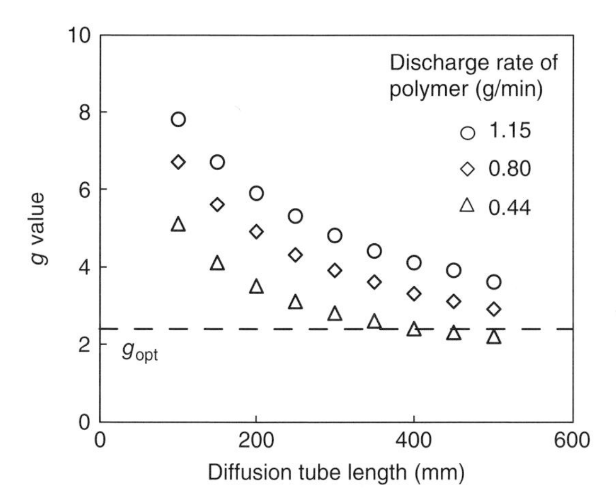

Figure 5.17 Changes in g values of poly(TFEMA)-DPS-based GI POF with length of the diffusion section. The diffusion temperature is assumed 180 °C and the initial core diameter 2.25 mm.

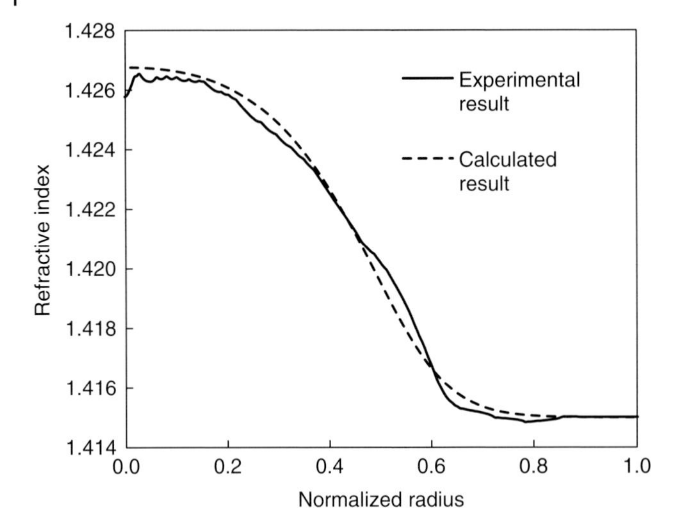

Figure 5.18 Comparison of experimental and calculated GI profiles. The fabrication and calculation conditions are as follows: diffusion temperature 180 °C; initial core diameter 2.25 mm; diffusion tube length 350 mm; and discharge rate of polymer 0.44 g/min.

by extrusion in accordance with the predicted parameters has nearly optimum profile and agrees with the simulated profile very well.

References

1. Kaino, T., Fujiki, M., and Nara, S. (1981) Low-loss polystyrene core-optical fibers. *J. Appl. Phys.*, **52** (12), 7061–7063.
2. Kaino, T., Jinguji, K., and Nara, S. (1983) Low loss poly(methylmethacrylate-d8) core optical fibers. *Appl. Phys. Lett.*, **42** (7), 567–569.
3. Mitsubishi Rayon Co. Ltd. (1974) Light transmitting fibers and method for making same. UK Patent 1,431,157, filed Jun. 20, 1974 and issued Apr. 7, 1976.
4. Ohtsuka, Y. and Nakamoto, I. (1976) Light-focusing plastic rod prepared by photocopolymerization of methacrylic esters with vinyl benzoates. *Appl. Phys. Lett.*, **29** (9), 559–561.
5. Ohtsuka, Y., Koike, Y., and Yamazaki, H. (1981) Studies on the light-focusing plastic rod. 6: the photocopolymer rod of methyl methacrylate with vinyl benzoate. *Appl. Opt.*, **20** (2), 280–285.
6. Ohtsuka, Y., Koike, Y., and Yamazaki, H. (1981) Studies on the light-focusing plastic rod. 10: a light-focusing plastic fiber of methyl methacrylate-vinyl benzoate copolymer. *Appl. Opt.*, **20** (15), 2726–2730.
7. Koike, Y., Takezawa, Y., and Ohtsuka, Y. (1988) New interfacial-gel copolymerization technique for steric GRIN polymer optical waveguides and lens arrays. *Appl. Opt.*, **27** (3), 486–491.
8. Koike, Y. (1991) High-bandwidth graded-index polymer optical fibre. *Polymer*, **32** (10), 1737–1745.
9. Debye, P. and Bueche, A.M. (1949) Scattering by an inhomogeneous solid. *J. Appl. Phys.*, **20** (6), 518–525.
10. Debye, P., Anderson, H.R., and Brumberger, H. (1957) Scattering by an inhomogeneous solid. II. The correlation function and its application. *J. Appl. Phys.*, **28** (6), 679–683.
11. Koike, Y. (1996) Optical resin materials with distributed refractive index, process for producing the materials, and optical conductors using the materials.

US Patent 5,541,247, filed Jun. 17, 1993 and issued Jul. 30, 1996.

12. Koike, K., Kado, T., Satoh, Z., Okamoto, Y., and Koike, Y. (2010) Optical and thermal properties of methyl methacrylate and pentafluorophenyl methacrylate copolymer: design of copolymers for low-loss optical fibers for gigabit in-home communications. *Polymer*, **51** (6), 1377–1385.
13. Van Duijnhoven, F.G.H. and Bastiaansen, C. (1999) Gradient refractive index polymers produced in a centrifugal field. *Adv. Mater.*, **11** (7), 567–570.
14. Van Duijnhoven, F.G.H. and Bastiaansen, C. (1999) Monomers and polymers in a centrifugal field: a new method to produce refractive-index gradients in polymers. *Appl. Opt.*, **38** (6), 1008–1014.
15. Im, S.H., Suh, D.J., Park, O.O., Cho, H., Choi, J.S., Park, J.K., and Hwang, J.T. (2002) Fabrication of a graded-index polymer optical fiber preform by using a centrifugal force. *Korean J. Chem. Eng.*, **19** (3), 505–509.
16. Im, S.H., Suh, D.J., Park, O.O., Cho, H., Choi, J.S., Park, J.K., and Hwang, J.T. (2002) Fabrication of a graded-index polymer optical fiber preform without a cavity by inclusion of an additional monomer under a centrifugal force field. *Appl. Opt.*, **41** (10), 1858–1863.
17. Shin, B.G., Park, J.H., and Kim, J.J. (2003) Low-loss, high-bandwidth graded-index plastic optical fiber fabricated by the centrifugal deposition method. *Appl. Phys. Lett.*, **82** (26), 4645–4647.
18. Koike, Y. (1997) Method of manufacturing plastic optical transmission medium. US Patent 5,593,621, filed Apr. 14, 1994 and issued Apr. 14, 1997.
19. Tanaka, T., Kurashima, K., Naritomi, M., Kondo, A., and Koike, Y. (2008) The first low-loss and high-bandwidth 61–127 channel graded-index steric cores polymer waveguide. Optical Fiber Communication Conference, San Diego, CA, February 24–28, 2008, p. OWG6.
20. Mukawa, Y., Kondo, A., and Koike, Y. (2012) Optimization of the refractive-index distribution of graded-index polymer optical fiber by the diffusion-assisted fabrication process. *Appl. Phys. Express*, **5** (4), 042501-1–3042501-3.
21. Asai, M., Hirose, R., Kondo, A., and Koike, Y. (2007) High-bandwidth graded-index plastic optical fiber by the dopant diffusion coextrusion process. *J. Lightwave Technol.*, **25** (10), 3062–3067.

6
Characterization

Characterization of optical fibers is very important in installing or designing optical fiber communication systems and improving their performance. Numerous measurement methods have been developed to evaluate the various performance indicators or parameters of optical fibers. This chapter explains several such methods selected for the characterization of the optical fibers described in this book and introduces methods of analyzing the obtained data by comparing the experimental results of several representative plastic optical fibers (POFs).

6.1
Refractive Index Profile

The bandwidth of an optical fiber strongly depends on the modal dispersion. The modal dispersion can be controlled by controlling the refractive index profile (RIP), which has a decisive effect on the bandwidth (see Chapter 3) and is a basic and important parameter of optical fibers. The pulse-broadening caused by modal dispersion seriously limits the transmission data rate of multimode fibers (MMFs) because overlapping of the broadened pulses induces intersymbol interference and disturbs correct signal detections, increasing the bit error rate [1].

6.1.1
Power-Law Approximation

A power-law index profile approximation is a well-known method of analyzing the RIP of graded-index (GI) MMFs [2]. In this approximation, the refractive index distribution of a GI MMF is described by

$$n(R) = \begin{cases} n_1\left[1 - 2\Delta\left(\frac{R}{a}\right)^g\right]^{\frac{1}{2}} & \text{for} \quad 0 \leq R \leq a \\ n_2 & \text{for} \quad R > a \end{cases}, \tag{6.1}$$

where $n(R)$ is the refractive index as a function of the radial distance R from the core center, n_1 and n_2 are the refractive indices of the core center and the cladding, respectively, and a is the core radius. The profile exponent g determines the shape

Fundamentals of Plastic Optical Fibers, First Edition. Yasuhiro Koike.

of the RIP, and Δ is the relative index difference, given by

$$\Delta = \frac{n_1^2 - n_2^2}{2n_1^2}. \tag{6.2}$$

Equation 6.1 includes the step-index (SI) profile when $g = \infty$.

The numerical aperture (NA) is another important parameter of an optical fiber. NA is related to the maximum angle at which light can be confined within the core by Snell's law and to the maximum number of propagating modes. The NA of the fiber as a function of the radial position from the core center is defined as [3]

$$\mathrm{NA}(R) = \begin{cases} \sqrt{n^2(R) - n_2^2} & \text{for} \quad 0 \le R \le a \\ 0 & \text{for} \quad R > a \end{cases}. \tag{6.3}$$

In SI MMFs, $n(R) = n_1$, so the NA is constant in the entire core region. In GI MMFs, the NA depends on the radial position and is usually described by the NA value at the core center.

6.1.2
Transverse Interference Technique

There are many methods to measure the RIP of the preform and the fiber. However, the RIP might be changed by the heat-drawing process, but some fabrication methods, such as coextrusion, can produce optical fibers directly from raw materials, not via preforms (see Chapter 5). In addition, we are interested in the final RIP of the fiber and not in the RIP of the preform during fabrication. Thus, this section focuses on measurements of the RIP of the fiber.

A transverse interferometric technique using interference microscopy is adopted to measure the RIP of the POF because of its high accuracy, high resolution, and ease of sample preparation. Figure 6.1 shows the principle of the technique, which assumes that the fiber has a rotationally symmetric structure around its axis. Here, only the principle of the technique is discussed, as the detailed mathematical explanations are given elsewhere [4, 5].

A fiber immersed in an oil with a refractive index close to that of the fiber cladding is observed transversely to its axis under the objective lens. The light beam through the fiber is divided into two paths by a beam splitter. One beam is displaced at a shearing distance Δy_{p} perpendicular to the light axis and then recombined with the other beam. When the Δy_{p} value is larger than the fiber diameter (total splitting), the two individual images are completely separated. Thus, an ordinary Mach–Zehnder interference pattern is observed, as shown in Figure 6.2a. When the Δy_{p} value is much smaller than the fiber diameter (partial splitting), an overlapping interference fringe is observed, as shown in Figure 6.2b. The shift difference ΔF of the fringe in partial splitting is related to the fringe shift F in total splitting as

$$\Delta F(y_{\mathrm{p}}) = F\left(y_{\mathrm{p}} + \frac{1}{2}\Delta y_{\mathrm{p}}\right) - F\left(y_{\mathrm{p}} - \frac{1}{2}\Delta y_{\mathrm{p}}\right), \tag{6.4}$$

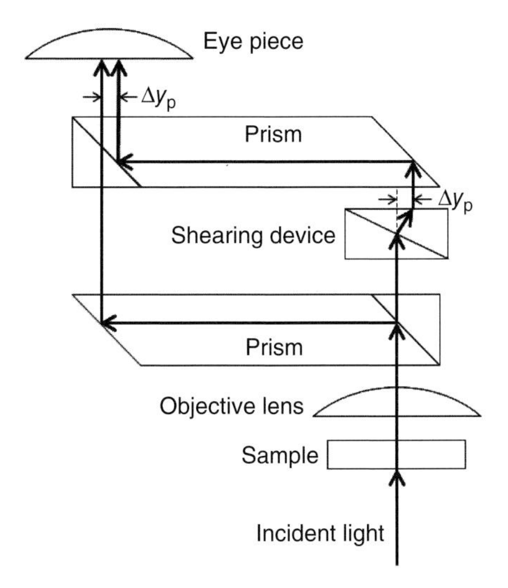

Figure 6.1 Schematic representation of transverse interferometric technique.

where y_p is the distance from the symmetry point of the fringe in partial splitting. If total splitting is applied for a relatively large sample, measurement of the fringe shift F is difficult because the fringe curve is significantly steep in the peripheral region and lies beyond one field of view. Partial splitting is preferred for measurement of a sample with a large diameter because the shift difference of the fringe can be accurately measured in one field of view.

The ray trajectory traversing the GI fiber when the refractive index of the immersion oil is different from that of the fiber cladding is shown in Figure 6.3. On the basis of the light refraction caused by the RIP of the fiber, the theoretical fringe shape is expressed as follows: an incident ray A is refracted with an angle θ and is focused on point B at a distance $y \sec \theta$ on the image plane. In total splitting, the fringe shift $F(y \sec \theta)$ at $y \sec \theta$ on the y_t-axis is described by

$$\frac{\lambda}{D} F(y \sec \theta) = \int_{P_0}^{P_1} n(R) ds - 2n_2 \sqrt{R_p^2 - y^2} - n_2 y \tan \theta, \tag{6.5}$$

where λ is the wavelength of the light source and D is the distance between adjacent interference fringes in the surrounding medium, which corresponds to one wavelength of the incident ray. The optical path length from P_0 to P_1 and the refraction angle $\theta(y)$ are expressed as

$$\int_{P_0}^{P_1} n(R) ds = 2n_p \sqrt{R_p^2 - (vy)^2} - 2 \int_{n_2 y}^{n_p R_p} \frac{d \ln n(u)}{du} \frac{u^2 du}{\sqrt{u^2 - (n_2 y)^2}}, \tag{6.6}$$

$$\theta(y) = \theta_0 + 2 \left[\sin^{-1} \left(\frac{y}{R_p} \right) - \sin^{-1} \left(\frac{vy}{R_p} \right) \right], \tag{6.7}$$

$$\theta_0 = -2n_2 y \int_{n_2 y}^{n_p R_p} \frac{d \ln n(u)}{du} \frac{du}{\sqrt{u^2 - (n_2 y)^2}}, \tag{6.8}$$

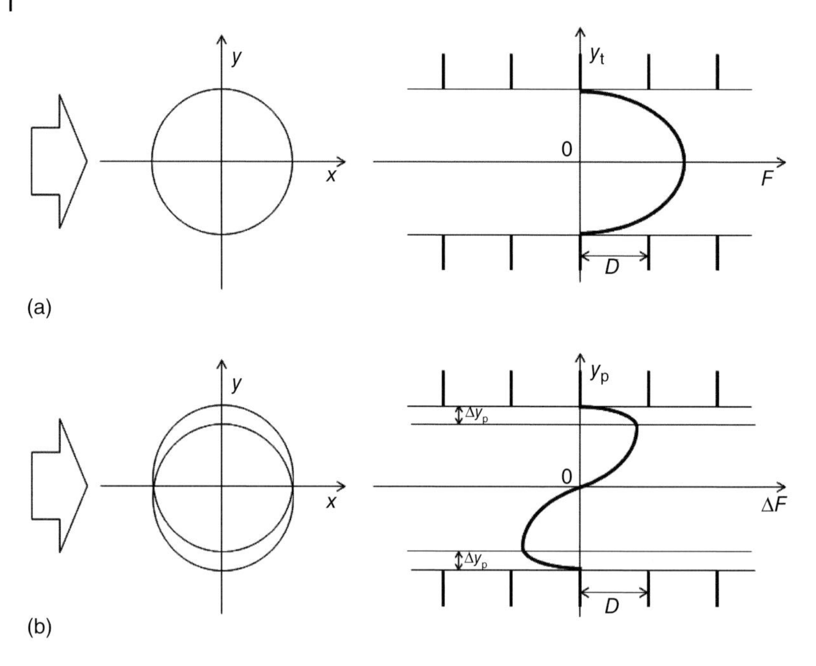

Figure 6.2 Interference fringes observed by transverse interferometric technique. (a) Total splitting. (b) Partial splitting.

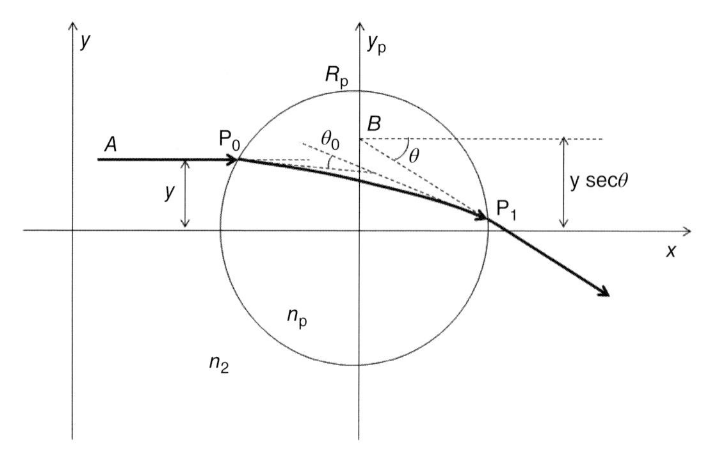

Figure 6.3 Ray trajectory traversing a GI fiber.

where $u \equiv Rn(R)$, $v \equiv n_2/n_p$, and θ_0 is the refraction angle inside the fiber. Thus, the transverse interferometric technique can determine the RIP accurately even when the immersion oil has a refractive index difference from that of the fiber cladding. This is a great advantage in practical use because it is quite difficult and time consuming to make an immersion oil with the same refractive index as the fiber cladding.

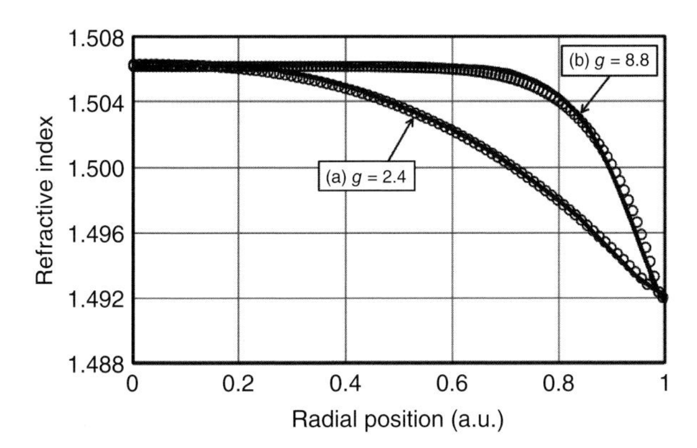

Figure 6.4 RIPs of GI POFs. (a) Almost ideal one ($g = 2.4$) and (b) one deviating greatly from the optimum ($g = 8.8$). Solid line: calculated from interference fringes measured by transverse interferometric technique; open circles: approximated by power-law equation.

Figure 6.4 shows representative RIPs of two poly(methyl methacrylate) (PMMA)-based GI POFs calculated from the interference fringes measured by the transverse interferometric technique. Both GI POFs were fabricated by interfacial-gel polymerization, and diphenyl sulfide was selected as the dopant to form the GI profile. In this case, for the optimum RIP, g_{opt}, is 2.4. Various RIPs for GI POFs, from the almost ideal to that deviating greatly from the optimum, can be precisely controlled by changing the polymerization rate of the interfacial-gel polymerization process. The abrupt (almost stepwise) change in the refractive index near the core–cladding boundary in the GI POF with $g = 8.8$ can be accurately measured, as well as the moderate change in the GI POF with $g = 2.4$. In addition, the calculated NAs had almost the same value of 0.205.

6.2
Launching Condition

The launching condition significantly influences important performance indicators, such as the attenuation and bandwidth, of an MMF in the absence of mode coupling as well as the optical power coupled into a fiber. This is because the launching condition determines the mode power distributions of MMFs, and different modes generally have different characteristics in MMFs. Therefore, the launching condition selected for the characterization of MMFs must be clarified. The launching condition is usually defined in terms of the properties of the optical signal injected from a light source into an optical fiber, that is, the NA, spot size, angle, and position of the incident beam.

In silica MMFs, a steady-state mode power distribution had been established in many previous studies because silica MMFs were expected to be used in long-haul networks. A steady-state mode power distribution was actually established in MMFs during real use [6]. However, MMFs, especially POFs, are currently expected to be the transmission medium in high-speed and short-reach networks, such as local area networks (LANs) in homes, office buildings, hospitals, cars, and airplanes. In such optical links, a laser diode (LD) or vertical-cavity surface-emitting laser (VCSEL) is a promising candidate for a light source in practical use. Because LDs or VCSELs generally have a very small radiation spot, only restricted small mode groups of the MMF are excited in such applications [7].

6.2.1 Underfilled and Overfilled Launching

The experimental setups for the launching conditions in some measurements are schematically illustrated in Figure 6.5. Two types of commercially available optical fibers with very short lengths are adopted as probe fibers for the underfilled launching (UFL) and overfilled launching (OFL) conditions. In both conditions, an optical beam from the LD is first injected into a very short probe fiber, and then the output light from the probe fiber is coupled to the test fiber by directly butting them. The probe fiber is located at the position where the core centers and fiber axes of both fibers coincide. The only difference between the two launching conditions is the probe fiber.

In the UFL condition, a silica-based MMF is selected as the probe fiber to excite restricted small mode groups. The core size and NA of the silica-based MMF used were 50 μm and 0.12, respectively. Because GI POFs typically have core diameters of 120–500 μm and NAs of less than 0.3, only small groups of lower order modes are selectively excited.

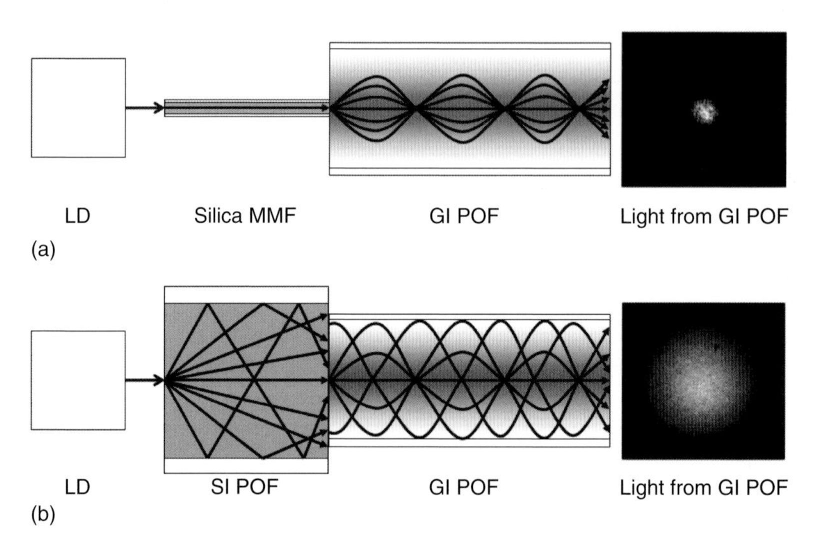

Figure 6.5 Schematic diagrams of (a) underfilled launching condition where only lower-order modes propagating near the core center are selectively excited. (b) Overfilled launching condition where all the modes are uniformly excited.

In the OFL condition, a very short SI POF is used as the probe fiber to uniformly excite all the modes. The core diameter and NA of the SI POF used were 980 μm and 0.5, respectively, which are much larger than those of typical GI POFs. In addition, the mode power distribution from even a short SI POF is uniform in the entire core region. Therefore, all the modes of the GI POF are uniformly excited [8].

6.2.2 Differential Mode Launching

The characteristics of the differential modes must be evaluated to analyze and improve the performance of MMFs because many important characteristics depend on the modes. Different modes have different characteristics in the absence of mode coupling. A very short silica-based single-mode fiber (SMF), instead of the MMF in Figure 6.5a, is used as the probe fiber to excite specified small mode groups. The core size and NA of the silica-based SMF used were 4.0 μm and 0.12, respectively. Small mode groups with similar propagation constants are selectively launched [9]. The position and angle of the incident beam are associated with the mode number, that is, the propagation constant, by

$$\frac{m}{m_{\max}} = \left[\left(\frac{R}{a}\right)^{g} + \frac{\sin^2\theta}{\sin^2\theta_{co}}\right]^{\frac{g+2}{2g}}, \tag{6.9}$$

where m is the principal mode number, $m_{\max}$ is the maximum principal mode number, R and θ are the radial position from the core center and the angle to the fiber axis of the incident beam, respectively, a is the core radius, and θ_{co} is the cut-off angle. Here, because the incident beam is aligned parallel to the fiber axis under test, the launched principal mode number is related only to the radial position of the incident light. Thus, under the differential mode launching condition (DML), small mode groups from lower order modes to higher order modes can be launched separately by shifting the position of the SMF from the core center to the periphery of the GI POF. In typical SI POFs, strong mode couplings are observed; hence, analyses of the characteristics of differential modes are insignificant. In earlier silica-based MMFs, the launching position was deliberately displaced from the core center in the well-known offset launching condition because early silica-based MMFs have a central dip in the RIP, which significantly reduces the attenuation and bandwidth.

6.3 Attenuation

Attenuation, or transmission loss, is one of the most important performance indicators of optical fibers. Attenuation is the main determinant of the maximum transmission distance of optical communication systems without amplifiers or repeaters, as well as of the maximum output power from the light source and the minimum receiver sensitivity. Attenuation is essentially caused by absorption, scattering, and radiation of optical power. These effects depend on both the intrinsic and extrinsic factors of the optical fibers. The optical power loss due to the individual contributions (see Chapter 2) is valuable information for researchers and manufacturers of fibers. On the other hand, the total attenuation is more interesting for system designers and users. Attenuation is defined as the ratio of the input and output powers. The optical power decreases exponentially with distance as light travels along a fiber. Therefore, the attenuation is discussed for a given fiber length and is typically described in decibels per kilometer (dB/km):

$$\alpha = -\frac{10}{L}\log_{10}\left(\frac{P_{\text{out}}}{P_{\text{in}}}\right), \tag{6.10}$$

where α is the total attenuation, P_{in} is the input optical power into the fiber, P_{out} is the output optical power from the fiber, and L is the fiber length. This section introduces the procedures for measuring the total attenuation.

6.3.1 Cutback Technique

The total attenuation can be determined by a cutback technique [10]. The cutback technique is the most widely used – but unfortunately destructive – method of measuring the attenuation because it is quite simple and fairly accurate. Figure 6.6 schematically illustrates an experimental setup for measuring the attenuation by the cutback technique. The optical power is first measured at the output end of the long fiber, and then the fiber is cut off a few meters from the output end, and the output power is again measured at this output end of the short fiber. This is because it is extremely difficult to detect directly the input power into the fiber. In the cutback technique, the output power from the short fiber is taken as the input power to a fiber with a length L corresponding to the distance between the two measurement points. In practice, the fiber is cut several times to improve the measurement accuracy. The launching condition (see Section 6.2) must be maintained identical throughout the measurement because different launching conditions produce different input optical powers to the fiber and different mode power distributions, resulting in different output powers from the fiber. In addition, the attenuation value strongly depends on the launching conditions, especially in MMFs with no or little mode coupling, because the modes of an MMF do not all have the same attenuation coefficient. The higher order modes typically

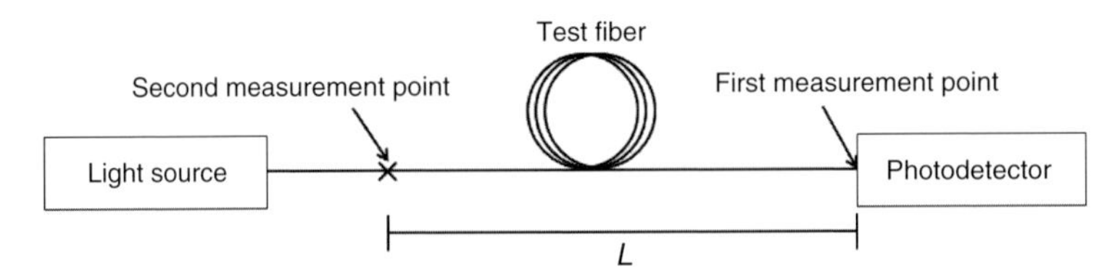

Figure 6.6 Attenuation measurement by the cutback technique.

exhibit higher attenuation coefficients in MMFs without central dips. The higher attenuation of higher order modes is generally explained by the effects of bending and the cladding. If the optical power is concentrated near the core center under the UFL, as shown in Figure 6.5, the attenuation contribution from higher order modes is negligible. In contrast, if the incident beam size and NA are larger than those of the test fiber under the OFL, the optical power of those parts of the incident light is lost. In addition, higher order modes contribute greatly to the attenuation. Thus, the OFL causes higher attenuation. Therefore, a mode scrambler, such as several turns of a “figure-of-eight” fiber bent around two mandrels with a slight separation and small radii of curvature, is typically inserted near the light source before the second measurement point in Figure 6.6 if steady-state equilibrium mode distributions are required.

Figure 6.7 shows the attenuation spectra of several GI POFs based on different polymers – PMMA, perdeuterated PMMA (PMMA-d_8), and perfluorinated polymer (which is commercially available from Asahi Glass Co., Ltd. under the trade name CYTOP) measured by the cutback technique under the OFL with a mode scrambler. Here, a white lamp with good temporal stability and a spectrum analyzer were used as the light source and photodetector, respectively. A light source or photodetector with a specific wavelength can be used if only the attenuation value at that wavelength is of interest. Valuable information, that is, the wavelengths and magnitudes of a low-loss window or absorption peak, is also obtained by measurement over a range of wavelengths, as shown in Figure 6.7. By substituting deuterium or fluorine for hydrogen, the absorption peak due to the vibration of the C–X bond is shifted to longer wavelengths, and the absorption intensity at a wavelength of interest is dramatically decreased [11–13]. The effects of deuteration and fluorination on the reduction of the attenuation are clearly observed. In particular, the CYTOP-based GI POF shows extremely low attenuation over a wide wavelength range. Therefore, CYTOP-based GI POFs can be used not only

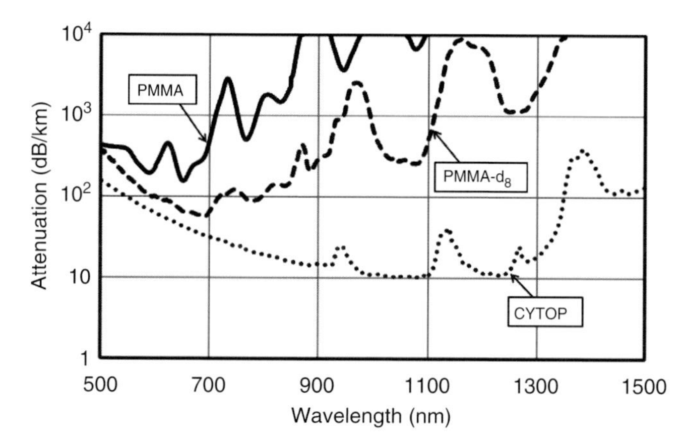

Figure 6.7 Attenuation spectra of GI POFs based on several polymers measured by the cutback technique.

in short-reach networks but also in networks spanning hundreds of meters, and also with light sources of various wavelengths.

6.3.2 Differential Mode Attenuation

Knowledge of which mode groups contribute to high attenuation is quite useful for improving the fibers. The attenuation of particular mode groups is obtained by measuring the differential mode attenuation (DMA). In addition, DMA clarifies the intrinsic material and waveguide properties [14]. The attenuation of lower order modes may be affected by the presence of a central dip or irregularities in the RIP, which were often observed in previous silica-based MMFs, or by a higher concentration of the dopant near the core center. Some dopants with a higher refractive index than the host material, which is necessary to form the GI profile, produce greater light scattering than the host material. The attenuation of high-order modes may be influenced by the cladding or primary coating and even the jacket of the fiber cable, or by structural imperfections at the core–cladding interface.

DMA is measured by the cutback technique under the DML (see Section 6.2). The output optical power from a long fiber is measured while shifting the position of the fiber probe over the input end face; then the fiber is cut off a few meters from the input end, and the output power is again measured at this output end of the short fiber while shifting the position of the fiber probe. The attenuation of the individual mode is calculated using Equation 6.10.

Figure 6.8 shows the DMA of a GI POF fabricated by interfacial-gel polymerization and another with double cladding fabricated by coextrusion (see Chapter 5). The attenuation of the GI POF fabricated by coextrusion increases rapidly as the launching position shifts from the core center to the core–cladding boundary, whereas a slight increment in the attenuation is observed in the GI POF fabricated by interfacial-gel polymerization. This is because the structural imperfections at the core–cladding interface that are formed during the coextrusion

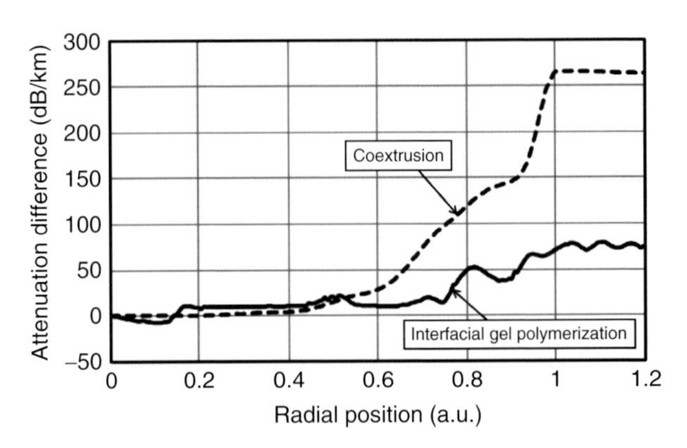

Figure 6.8 DMA of GI POFs fabricated by interfacial-gel polymerization and by coextrusion.

process induce excess light scattering and result in high attenuation [15, 16]. The launching position corresponds to the principal mode number in the GI POF (see Section 6.2). Thus, the DMA strongly depends on, and can be controlled by, the fabrication method.

6.4 Bandwidth

Bandwidth is one of the most important parameters of optical fibers, in addition to attenuation. Bandwidth determines the maximum transmission data rate or the maximum transmission distance. Most optical fiber communication systems adopt pulse modulation. If an input pulse waveform can be detected without distortion at the other end of the fiber, except for a decrease in the optical power, the maximum link length is limited by the fiber attenuation. However, in addition to the optical power attenuation, the output pulse will also be generally broader in time than the input pulse. This pulse broadening restricts the transmission capacity, namely, the bandwidth of the fiber. The bandwidth is determined by the impulse response as follows [17]: Optical fibers are usually considered quasi-linear systems, and therefore the output pulse is described by

$$p_{\text{out}}(t) = h(t) * p_{\text{in}}(t). \tag{6.11}$$

The output pulse $p_{\text{out}}(t)$ from the fiber can be calculated in the time domain through the convolution (denoted by *) of the input pulse $p_{\text{in}}(t)$ and the impulse response function $h(t)$ of the fiber. Fourier transformation of Equation 6.11 provides a simple expression as the product in the frequency domain:

$$P_{\text{out}}(f) = H(f)P_{\text{in}}(f), \tag{6.12}$$

where $H(f)$ is the power transfer function of the fiber at the baseband frequency f [18]. The power transfer function defines the bandwidth of the optical fiber as the lowest frequency at which $H(f)$ is reduced to half its DC value. The bandwidth of the fiber is often called the *−3-dB bandwidth* because the half-value is −3 dB. The power transfer function is easily calculated from the Fourier transform of the experimentally measured input and output pulses in the time domain, or from the measured output power from the fiber in the frequency domain from DC to the bandwidth frequency.

A higher bandwidth yields less pulse broadening and enables higher speed data transmission. The bandwidth limitation also largely determines the maximum link length for a given data rate in some MMF systems. Pulse broadening, which is theoretically proportional to the fiber length, is caused mainly by two dispersion mechanisms in POFs: intermodal and intramodal. Another type of dispersion, polarization mode dispersion, arises from the nonuniformity of the structure and material throughout the fiber, whereby orthogonal polarization modes are affected by the slightly different refractive indices [19]. The effect of polarization mode dispersion can usually be ignored in MMFs. Researchers and manufacturers are interested in all dispersion mechanisms, whereas system designers and users care more about the bandwidth. This section explains the method of measuring the fiber bandwidth using a time domain method.

6.4.1 Time Domain Measurement

An ideal bandwidth measurement in the time domain gives the impulse response of a fiber directly. However, direct measurement of the impulse response is impractical because an ideal delta function signal is not available. Thus, a simple procedure for bandwidth measurement is illustrated in Figure 6.9. A narrow light pulse from a light source, usually a laser, is injected into the test fiber, and the waveform of the broadened output pulse from the fiber is detected and recorded by an optical sampling oscilloscope. The output pulse waveform from a short reference fiber, regarded as the input pulse waveform into the test fiber, is also measured in the same way. An electrical trigger signal with a suitable delay for the fiber length is used to detect the optical pulse with the oscilloscope. The power transfer function of the test fiber is calculated by Equation 6.12 from the Fourier transformation of the measured input and output pulse waveforms. The bandwidth of the fiber is determined by the lowest frequency at which the obtained power transfer function is reduced to half its maximum value. Because the bandwidth of an MMF strongly depends on the launching condition, the condition adopted in the measurement must be specified.

Figure 6.10 shows the input and output pulse waveforms of representative PMMA-based GI POFs with a core diameter of 500 μm and the RIPs shown in Figure 6.4 under the UFL compared with those of a PMMA-based low-NA (0.3) SI POF. In addition, those of the GI POF with the nearly optimum RIP under the OFL are also shown. The low-NA SI POF shows a dramatically broadened output pulse and a narrow bandwidth of 240 MHz for 50 m, although strong mode coupling in SI POFs generally improves the bandwidth. A bandwidth of 230 MHz for 100 m was obtained even in the GI POF with a far-from-optimum RIP, which is almost the same as that of the SI POF even though the length of the GI POF was twice that of the SI POF. On the other hand, the GI POF with the almost optimized RIP exhibits an output pulse almost the same as the input pulse, with almost no broadening in time and a high bandwidth of 2.6 GHz for 100 m.

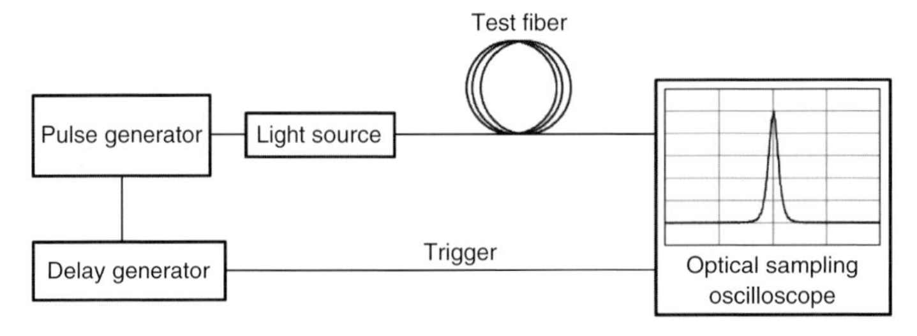

Figure 6.9 Bandwidth measurement in the time domain.

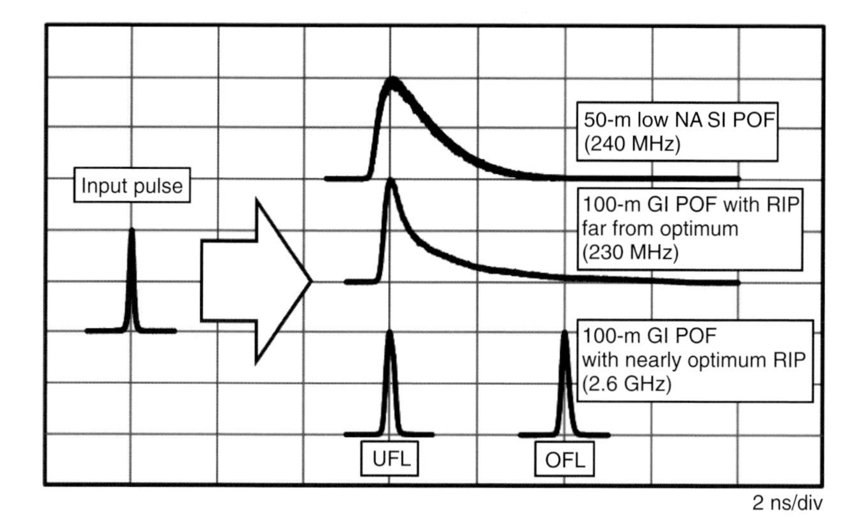

Figure 6.10 Input and output pulse waveforms of 100-m GI POFs with the RIPs shown in Figure 6.4 under UFL compared with that of 50-m low-NA SI POF.

Furthermore, the bandwidth even under the OFL, which is considered one of the worst cases, was as high as that under the UFL. Thus, the almost optimum GI POF has high bandwidth characteristics independent of the launching condition because all the modes of the GI POF with nearly the optimum RIP have almost the same delay time. These results, of course, might be due to strong mode coupling; however, the following sections will demonstrate that mode coupling has little effect. The difference in the bandwidths of the various POFs described above is due mainly to modal dispersion (see Chapter 3).

6.4.2 Differential Mode Delay

Knowledge of the differential mode delay (DMD) is extremely valuable for assessing and improving the performance of MMFs because departures from the ideal RIP are more easily confirmed by observing the DMDs than by measuring the RIPs. DMD measurements provide information about the RIP deviations averaged over the fiber length. Of course, DMD measurements do not replace RIP measurements, but rather complement them. DMD measurement is limited in the absence of significant mode coupling. The time delay of a given mode group can be measured only if the light power remains in the mode group that is initially excited. If the power is redistributed to other mode groups by mode coupling, DMD measurement becomes meaningless. When strong mode coupling exists, the radiation pattern of light from the far end of the fiber becomes independent (or nearly so) of the launching condition (see the next section).

Different mode groups can be selectively excited in an SMF by scanning the launching position under the DML, as shown in Figure 6.11. The DMD is evaluated

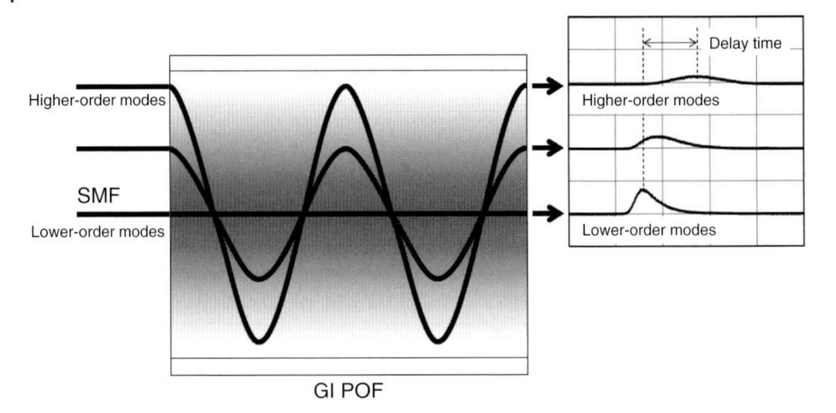

Figure 6.11 Schematic diagram of DMD measurement.

by measuring the output pulse waveforms and the delay times of the pulses as a function of the radial position of the fiber probe or, equivalently, as a function of the normalized principal mode number in time domain measurement. Different modes generally have different delay times in the absence of mode coupling.

The GI POF with the nearly optimum profile exhibited high bandwidth and almost the same bandwidth under the UFL and OFL, independent of the launching condition, as mentioned in the previous section. Strong mode coupling is generally suspected in such cases. Thus, the effect of mode coupling was evaluated by measuring the DMD. Figure 6.12 shows the DMD of the GI POF with almost the optimum RIP shown in Figure 6.4 compared to that of the GI POF with a nonideal g of 4.8 and a low NA of 0.15. The DMD of the GI POF with the nearly optimum RIP clarified that all the modes have almost the same delay times. Therefore, the optimum GI POF exhibited high bandwidth independent of the launching condition. On the other hand, the nonideal GI POF also had almost the same DMDs, even though the RIP deviated from the optimum one. In addition, this nonideal GI POF exhibited a high bandwidth of 2.5 GHz. Therefore, strong mode coupling is the primary factor. The quite close delay times can be caused by the optimum RIP or might be affected by strong mode coupling. Because the results of the DMD measurements alone cannot rigorously determine which effect is dominant, a more detailed evaluation of mode coupling will be discussed in the next section.

6.5 Near-Field Pattern

A transmitted near-field pattern (NFP) measurement method is applied for measuring the mode field characteristics, which correspond to the radial optical field distribution across the core [12, 13]. NFP measurement is typically used to determine the fiber parameters, for example, the mode field diameter, core and cladding

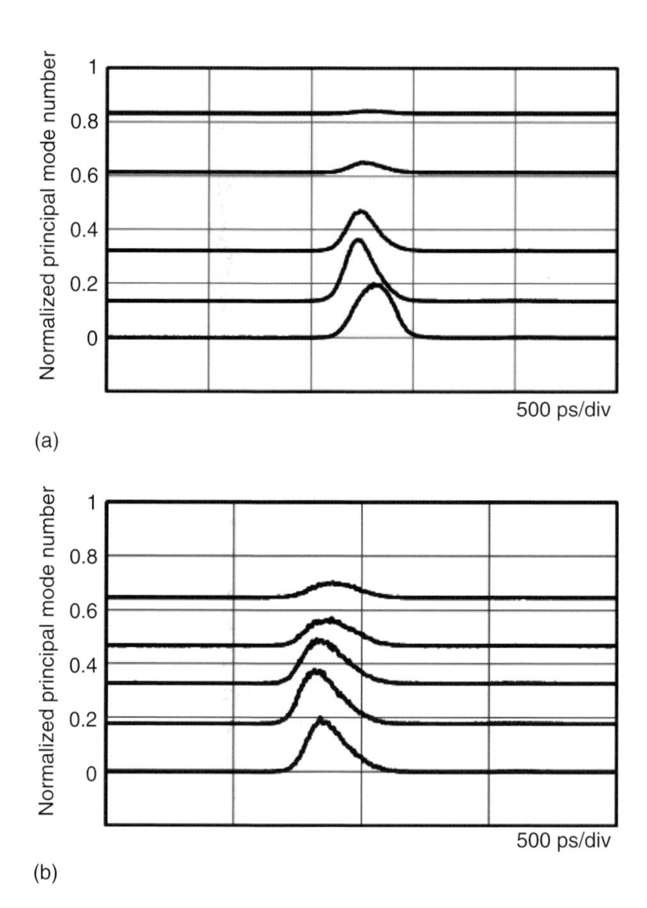

Figure 6.12 DMDs of GI POFs with (a) almost the optimum RIP and (b) nonideal RIP and low NA.

diameters, concentricity, circularity, and RIP [14]. The mode field is very important, especially in SMFs, because it affects various characteristics such as the coupling efficiency and dispersion. On the other hand, the NFP directly provides the optical intensity distribution (equivalently, the mode power distribution) across the output end face of the fiber. Thus, the NFP strongly depends on the launching condition in the absence of mode coupling.

The differential mode NFPs are measured by moving the position of the SMF under the DML. If there is no or only little mode coupling, the differential mode NFPs are apparently different from each other, as shown in Figure 6.13. In typical GI POFs with little mode coupling effect, the optical power, that is, the NFP, of the lower order modes is confined near the core center whereas a ring-shaped NFP of the higher order modes is observed.

The previous section mentioned two GI POFs, one with almost the optimum RIP and the other with a nonideal RIP and low NA, all modes of which had almost the same delay times. DMD measurements alone cannot determine whether

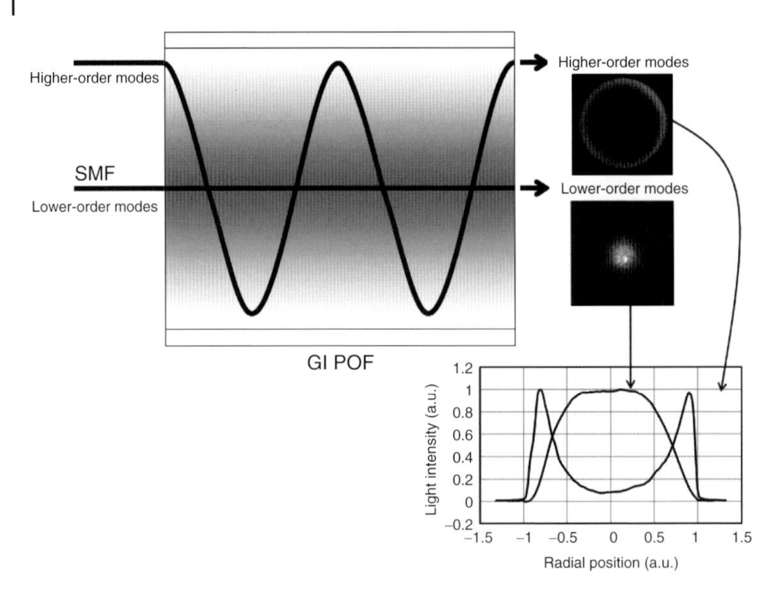

Figure 6.13 Representative differential mode NFPs of GI POFs.

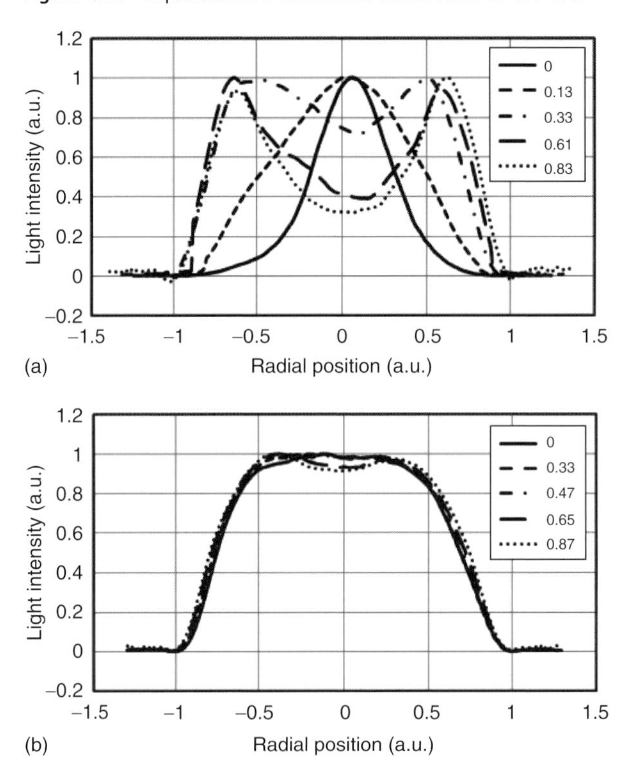

Figure 6.14 Differential mode NFPs of GI POFs with (a) almost the optimum RIP and (b) nonideal RIP and low NA. Legends indicate the normalized principal mode number.

the similar delays originate from the optimum RIP or strong mode coupling. Figure 6.14 shows the differential mode NFPs of 100-m lengths of these GI POFs. The nearly optimum GI POF had distinct NFPs corresponding to the excited mode groups, indicating that the mode coupling effect is quite small. In contrast, the nonideal GI POF exhibited similar NFPs independent of the launched mode groups because of strong mode coupling caused by the low NA of the GI POF [20, 21]. Consequently, the high bandwidths are attributed mainly to the RIP in the GI POF with almost the optimum RIP as well as to the strong mode coupling effect in the GI POF with a nonideal RIP and low NA.

References

1. Nowell, M.C., Cunningham, D.G., Hanson, D.C., and Kazovsky, L.G. (2000) Evaluation of Gb/s laser based fibre LAN links: review of the Gigabit Ethernet model. *Opt. Quantum Electron.*, **32** (2), 169–192.
2. Olshansky, R. (1979) Propagation in glass optical waveguides. *Rev. Mod. Phys.*, **51** (2), 341–367.
3. Gloge, D. and Marcatili, E.A.J. (1973) Multimode theory of graded-core fibers. *Bell Syst. Tech. J.*, **52** (9), 1563–1578.
4. Kokubun, Y. and Iga, K. (1977) Precise measurement of the refractive index profile of optical fibers by a nondestructive interference method. *Trans. IECE Jpn.*, **E60** (12), 702–707.
5. Ohtsuka, Y. and Koike, Y. (1980) Determination of the refractive-index profile of light-focusing rods: accuracy of a method using Interphako interference microscopy. *Appl. Opt.*, **19** (16), 2866–2872.
6. Marcuse, D. (1973) Coupled mode theory of round optical fibers. *Bell Syst. Tech. J.*, **52** (6), 817–842.
7. Raddatz, L., White, I., Cunningham, D., and Nowell, M. (1998) An experimental and theoretical study of the offset launch technique for the enhancement of the bandwidth of multimode fiber links. *J. Lightwave Technol.*, **16** (3), 324.
8. Marcuse, D. (1975) Excitation of parabolic-index fibers with incoherent sources. *Bell Syst. Tech. J.*, **54** (9), 1507–1530.
9. Jeunhomme, L. and Pocholle, J.P. (1978) Selective mode excitation of graded index optical fibers. *Appl. Opt.*, **17** (3), 463–468.
10. IEC (2001) Optical Fibres: Part 1-40: Measurement Methods and Test Procedures: Attenuation.
11. Koike, Y. and Naritomi, M. (1994) Graded-refractive-index optical plastic material and method for its production. JP Patent 3719733, US Patent 5783636, EU Patent 0710855, KR Patent 375581, CN Patent L951903152, TW Patent 090942, originally filed in 1994.
12. Nihei, E., Ishigure, T., and Koike, Y. (1996) High-bandwidth, graded-index polymer optical fiber for near-infrared use. *Appl. Opt.*, **35** (36), 7085–7090.
13. Makino, K., Kado, T., Inoue, A., and Koike, Y. (2012) Low loss graded index polymer optical fiber with high stability under damp heat conditions. *Opt. Express*, **20** (12), 12893–12898.
14. Olshansky, R. and Oaks, S.M. (1978) Differential mode attenuation measurements in graded-index fibers. *Appl. Opt.*, **17** (11), 1830–1835.
15. Noda, T. and Koike, Y. (2010) Bandwidth enhancement of graded index plastic optical fiber by control of differential mode attenuation. *Opt. Express*, **18** (3), 3128–3136.
16. Makino, K., Akimoto, Y., Koike, K., Kondo, A., Inoue, A., and Koike, Y. (2013) Low loss and high bandwidth polystyrene-based graded index polymer optical fiber. *J. Lightwave Technol.*, **31** (14), 2407.
17. Keiser, G. (2010) *Optical Fiber Communications*, 4th edn, McGraw-Hill.
18. Personick, S. (1973) Baseband linearity and equalization in fiber optic digital communication systems. *Bell Syst. Tech. J.*, **52** (7), 1175–1194.

19. Poole, C.D. and Wagner, R.E. (1986) Phenomenological approach to polarisation dispersion in long single-mode fibres. *Electron. Lett.*, **22** (19), 1029–1030.
20. Makino, K., Nakamura, T., Ishigure, T., and Koike, Y. (2005) Analysis of graded-index polymer optical fiber link performance under fiber bending. *J. Lightwave Technol.*, **23** (6), 2062–2072.
21. Makino, K., Ishigure, T., and Koike, Y. (2006) Waveguide parameter design of graded-index plastic optical fibers for bending-loss reduction. *J. Lightwave Technol.*, **24** (5), 2108–2114.

7
Optical Link Design

The basic elements of an optical link serve the same purpose as in other communication systems: to transport information from one location to another. The essential components consist of a transmitter, a receiver, and an optical fiber. Additional components such as amplifiers, regenerators, multiplexers, demultiplexers, filters, equalizers, connectors, and splices can be used if needed. The transmitter is a light source, usually a laser diode (LD) or a vertical-cavity surface-emitting laser (VCSEL) for high-speed data rates, which converts an electrical signal to an optical signal, acting as a so-called E/O converter. Both analog and digital signals can be applied. The receiver is a photodetector, usually a photodiode whose material depends on the signal wavelength, which converts an optical signal into an electrical signal, acting as a so-called O/E converter. Different optical fibers can be selected from a large variety of fibers depending on the application and environment. For example, a silica-based single-mode fiber (SMF) is suitable for a long-haul undersea network, whereas a plastic optical fiber (POF) is adequate for local area networks (LANs) in homes, offices, hospitals, vehicles, and airplanes.

An eye diagram provides important characteristics such as the bit error rate (BER), signal-to-noise ratio (SNR), and timing jitter for the design, operation, and maintenance of an optical link. These characteristics, of course, are parameters of the communication system; that is, they depend not only on the optical fibers but also on the optical devices and optoelectronic components, such as light sources, photodetectors, and amplifiers [1, 2]. In long-haul optical communication, for example an undersea optical fiber network using silica-based SMFs, accurate and comprehensive evaluations of the optical fiber are required because undersea optical fibers cannot be easily replaced. In addition, the characteristics of the components in communication systems depend on the operating conditions and environment, and different characteristics are required for different communication systems. The performance of the link depends on various factors, some of which affect each other. Link design obviously suffers from various types of tradeoff. This chapter theoretically and experimentally discusses the performance of optical links with fixed optoelectronic devices, focusing on optical fibers with different characteristics, on the basis of the link power budget model.

Fundamentals of Plastic Optical Fibers, First Edition. Yasuhiro Koike.

7.1 Link Power Budget

The power difference between the optical transmitter output and the minimum receiver sensitivity required to satisfy a specified BER is known as the *link power budget* and is crucial to the link design [3]. The maximum output power from the transmitter is limited by the receiver saturation or the maximum output power of the transmitter (or the eye safety in a consumer network). The link power budget is assigned to various losses and power penalties of the optical channel and to the safety link margin, as shown in Figure 7.1. The link losses include the optical power loss due to fiber attenuation, splices, and connectors. Power penalties denote the additional power required for the transmitter output to suppress the increases in the BER due to noise, interference, and impairment. The safety margin is demanded to ensure the link performance regardless of changes in the characteristics due to temperature fluctuations and degradation caused by component aging. The optical link design means the assignment of the power budget to all these factors.

7.2 Eye Diagram

An eye diagram is the most commonly used method of evaluating the performance of digital data communication systems and is also applied to optical fiber networks. The eye diagram provides valuable information on the optical link performance, such as the signal distortion, rise time, noise margin, and timing

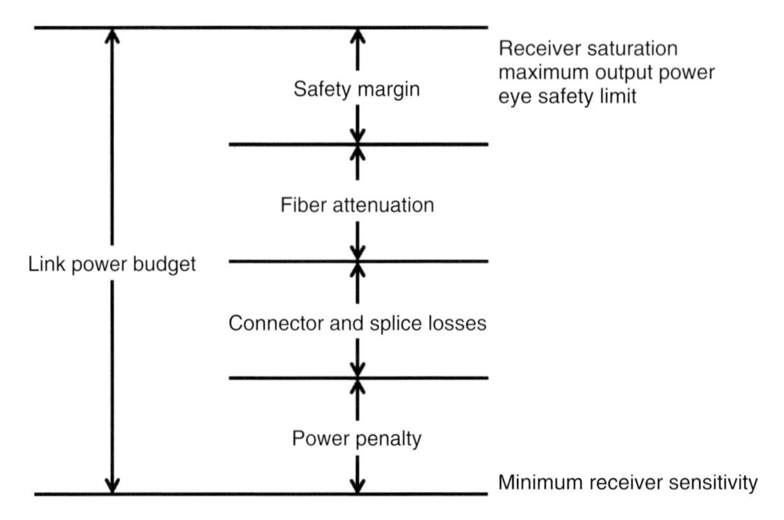

Figure 7.1 Concept of link power budget.

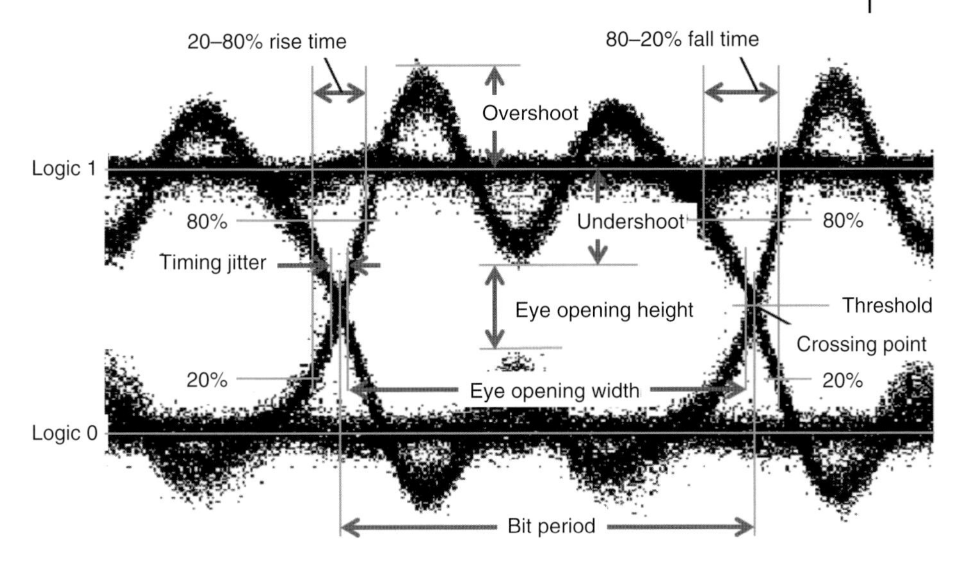

Figure 7.2 Fundamental parameters of eye diagram.

jitter [4]. The eye diagram can be observed in the time domain by generating a pseudorandom bit sequence of logic 0 and 1 at a uniform rate in a random pattern. When the bits of the pseudorandom sequence are superimposed simultaneously, a unique pattern is observed that looks like a human eye, as shown in Figure 7.2: the so-called eye diagram or eye pattern. Thus, the eye diagram is important for estimating the link performance in worst case transmission. The pseudorandom binary sequence (PRBS) repeatedly generates randomly different N-bit combinations. The length of the PRBS is generally described by $2^N - 1$, where N is an integer. Thus, the repetition rate of the PRBS is unrelated to the data rate.

7.2.1 Eye Opening

Figure 7.2 shows a typical eye diagram and defines the fundamental parameters for evaluating a communication system. The threshold is used to determine the logic value of the received bit. Although the threshold is defined by various methods, it is indicated by the crossing point here. The width of the eye opening defines the time interval during which no signal exists at the threshold level. The height of the eye opening at a specified sampling time shows the noise margin, which is the ratio of the height of the eye opening to the maximum signal from the threshold level. The best sampling time of the received signal is the point of the greatest height of the eye opening. The eye height is reduced by signal attenuation and distortion in the channel. A large height of the eye opening leads to a high probability of distinguishing correctly whether the signal is logic 0 or 1. The slopes of the eye diagram are equivalent to the tolerance of the sampling timing. If the slope becomes horizontal, the possibility of error decision due to the shift in the sampling time increases. The rise time is generally defined as the time period between the points where the signal intensity of logic 1 is 10% and 90%. Also, the time period between the 20% and 80% points is often adopted in optical links because these points can be more accurately measured. Timing jitter or phase distortion occurs when the signal changes from one symbol state to the other earlier or later than the exact end of the bit period. Timing jitter is quite important for high-speed communication systems because the pulses are very close to each other. In this case, incorrect interpretation of the bit edge leads to a high BER. Timing jitter in optical fiber transmission systems is attributed to various types of noise and pulse distortion due to dispersion and interchannel crosstalk. Excess jitter indicates an incorrect sampling timing and results in incorrect bit decisions.

7.2.2 Eye Mask

The eye diagram is generally distorted by signal deterioration because of the total attenuation and dispersion of the entire optical link, including the transmitter, the receiver, and the optical fiber. The received eye is more closed than the transmitted eye. Asymmetric eye patterns are also often observed in practice because of nonlinear effects in the channel. All the eye openings should be identical and symmetrical if a random data stream passes through a genuinely linear system. A distorted eye diagram is evaluated by an eye mask specified by a standard. If the received eye is wider than the mask, the data are guaranteed to satisfy the specified BER. The eye mask is defined by a polygon – typically a hexagon, rectangle, or diamond – located in the middle of the eye opening. The mask height corresponds to the signal power and represents the minimum difference between the data values of 0 and 1 required to satisfy a specific BER. The mask width is inversely proportional to the bit rate and is also related to the allowed timing jitter. The slopes of the polygon indicate the acceptable rise and fall times. Figure 7.3 shows the eye pattern and eye mask of a Gigabit Ethernet link consisting of a 50-m graded-index (GI) POF [5]. No signal was detected inside the eye mask, although a relatively large jitter was observed because the transmitter and the receiver were not optimized. Thus, the GI POF can ensure reliable data transmission in a Gigabit Ethernet link.

7.3 Bit Error Rate and Link Power Penalty

The average error probability is the most useful criterion for evaluating the performance of digital communication systems. A decision circuit samples the signal voltage at the midpoint or a determined sampling time of each time slot and

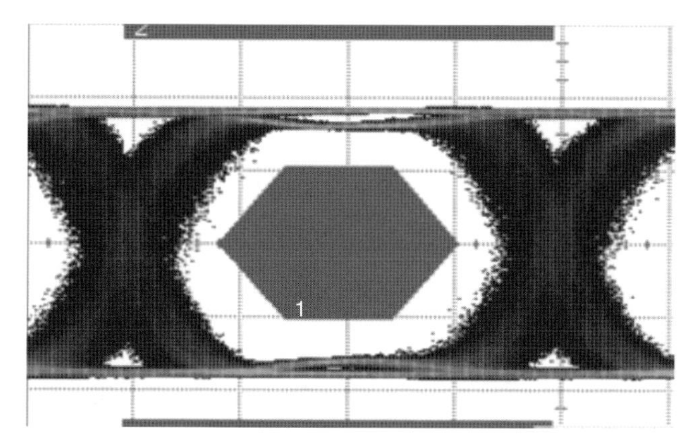

Figure 7.3 Eye mask of Gigabit Ethernet.

compares the value with a certain reference voltage known as the *threshold*. If the received signal is larger than the threshold, the circuit determines that the received bit is logic 1. If the signal is smaller than the threshold, the bit is logic 0. A clock signal, periodically equal to the bit interval, is used to register the start point of the bit and hence the sampling timing for the bit decision. The clock is typically encoded within a data signal. The clock must be extracted from the data stream before bit determination; thus, this function is called *clock recovery*. Ideally, the received signal always exceeds the threshold when logic 1 is sent and is always less than the threshold when logic 0 is sent.

However, in practical systems, the signal is degraded by various types of noise, interference from adjacent pulses, and the finite extinction ratio of the light source. If an optical transmission system has no optical amplifier, thermal and shot noises are the dominant factors in the receiver [6]. Thermal noise is independent of the incoming optical power; in contrast, shot noise depends on the received optical power. A critical error source is intersymbol interference (ISI), which arises from pulse spreading due to dispersion [7]. The pulse broadening causes the received signal to spread into the adjacent bit period. Thus, errors in the bit decision occur.

BER is simply defined as the number of bits received with errors divided by the total number of bits received within a certain time interval. An optical communication system specifies a BER value to ensure the link performance. Typical BERs required for optical fiber communication systems range from 10^{-9} to 10^{-12}. For example, Gigabit Ethernet requires a BER of no more than 10^{-12} [5]. On the other hand, precise equalizer and error correction techniques, such as forward error correction (FEC), have recently attracted a great deal of attention in extremely high speed networks [8]. In FEC, redundant data are transmitted within the original data stream. If some of the original data are lost or received incorrectly, the redundant data are used to reconstruct or complement them. Thus, the allowed BER is dramatically relaxed to around 10^{-3} by using FEC and/or an equalizer [9]. To satisfy a desired BER at a given data rate, a specific minimum average

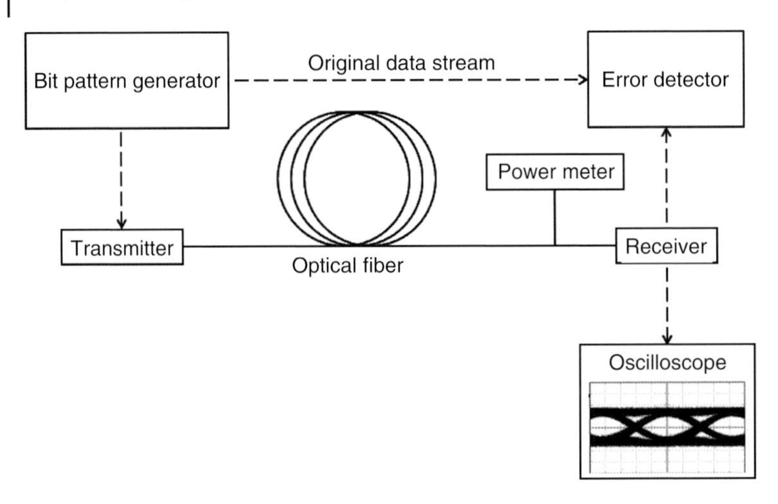

Figure 7.4 BER and eye diagram measurement apparatus.

optical power must be input to the photodetector. The value of the minimum power is known as the *minimum receiver sensitivity*. If an ideal photodetector that has unity quantum efficiency and produces no dark current, the minimum received optical power required for a specific BER in a digital system is called the *quantum limit*, because the link performance is limited only by the statistics of the photodetection mechanism [10]. Because the BER is a statistical parameter, the value depends on the measurement time as well as on the error factors. A BER of 10^{-12} in a 1 Gb/s network corresponds to only one bit error in every 1000 s, so around 28 h are required to detect 100 errors. Thus, the test times can be quite long. Several techniques have been developed to reduce such costly and time-consuming test periods. Although some accuracy is exchanged for the reduction in the measurement time, the required time can be dramatically decreased from hours to minutes. Figure 7.4 shows a typical BER and eye diagram measurement apparatus, which consists of a bit pattern generator, a transmission link or device under test, and a bit error detector associated with the pattern generator. A PRBS created by the pattern generator is transmitted through the optical transmission system. The error detector compares the received data stream with the original data stream. These two signal streams must be perfectly synchronized to be compared correctly in BER measurements. In addition, the received data stream is superimposed simultaneously and displayed on the oscilloscope as the eye diagram.

Various types of signal impairment that are inherent in optical fiber transmission systems degrade the link performance. Any signal impairment in an optical link reduces the optical power transmitted to the receiver compared to the ideal case. This lower power reduces the SNR of the link compared to the case with no impairment [11]. Because a reduced SNR leads to a higher BER, additional signal power is required at the receiver to maintain the same BER as in the ideal case. The required additional optical power is known as the *power penalty* and is generally expressed in decibels (dB). If there are nonlinear effects, the increase in the received optical power has no effect on the improvement of the BER [12]. In addition, the reflection of laser light from the optical fiber or the receiver back into the laser severely degrades the performance of the laser [13]. Because the laser is a noise source in the presence of strong back reflection, increasing the output optical power from the transmitter cannot improve the BER of the link.

7.3.1
Intersymbol Interference

ISI arises from pulse broadening due to various types of dispersion. When a pulse is transmitted in a given bit interval, most of the output pulse energy arrives at the receiver in the corresponding time slot. However, some of the transmitted energy spreads into adjacent bit periods because the pulse has spread. The pulse intruding from the adjacent bit period disturbs the correct bit decision of the original signal [14, 15]. Thus, ISI degrades the BER. The power penalty due to ISI (in linear units) is described by

$$P_{\mathrm{ISI}} = \frac{1}{1 - 1.425 \exp\left[-1.28\left(\frac{T}{T_{\mathrm{c}}}\right)^2\right]}, \tag{7.1}$$

where T is the bit period and T_{c} is the channel response time. The dispersion of the optical fiber is generally a dominant factor in ISI, especially in multimode fiber (MMF) systems, although the transmitter and receiver also affect the channel response time.

Figure 7.5 shows the measured BER curves of two 50-m GI POFs with an almost ideal refractive index profile (RIP) of $g = 2.4$ and with a nonideal RIP of $g = 6.4$ (see Chapter 3) compared to the back-to-back value, which is the link performance of the measurement system consisting of a very short (1 m) fiber in which the effect of fiber attenuation and dispersion is negligible. The data rate was set to 500 Mb/s, which corresponds to IEEE1394 S400. An LD with a wavelength of 650 nm and a PIN photodiode were used here. Both the GI POFs were based on the same material poly(methyl methacrylate), (PMMA) and have the same length, core diameter, and numerical aperture (NA) but different RIPs. Therefore, the only difference between the two GI POFs is in the modal dispersion. A serious power penalty due to ISI caused by the large modal dispersion is apparently observed in the GI POF with a nonideal RIP. In contrast, the nearly optimum GI POF exhibits almost the same BER curve as the system sensitivity; that is, no power penalty is practically caused by the GI POF. Therefore, the GI POF can realize reliable data transmission and provide a sufficient margin in the link design. Thus, because POFs generally have large modal dispersion, controlling such dispersion by means of the RIP is quite important for suppressing ISI and hence the BER degradation and power penalty. ISI is strictly restricted by various standards;

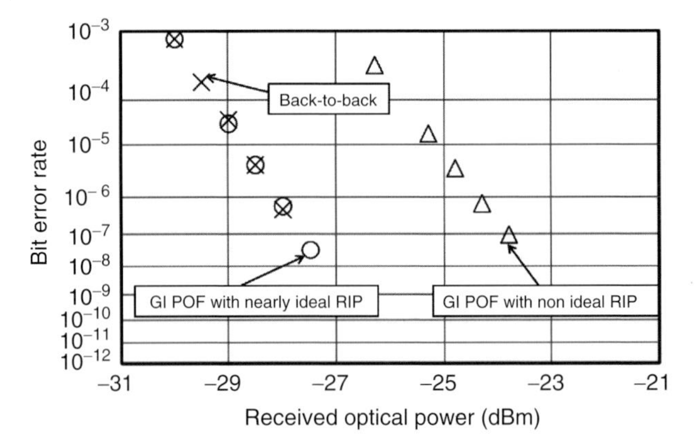

Figure 7.5 BER degradation due to ISI.

for example, Gigabit Ethernet specifies that the acceptable penalty due to ISI be within 3.6 dB [5].

7.3.2
Extinction Ratio

The extinction ratio of a laser in an optical transmission link is defined as the ratio of the optical power for a logic 1 signal to the power for a logic 0 signal. Ideally, the extinction ratio should be infinite. However, the ratio must be finite in an actual system in order to reduce the rise time of the laser pulses [16]. The nonzero power transmission for a bit of logic 0 causes the power penalty. The power penalty due to the extinction ratio (in linear units) is estimated by

$$P_{\mathrm{ER}} = \frac{1+\epsilon}{1-\epsilon}, \tag{7.2}$$

where ϵ is the laser extinction ratio.

7.3.3
Mode Partition Noise

Mode partition noise is associated with intensity fluctuations among the longitudinal modes of a multimode laser, such as an edge-emitting laser. This is the dominant noise in SMF systems when multimode lasers are used. In a multimode laser, the power in each mode is not constant even when the total power is constant [17, 18]. Different longitudinal modes have different attenuation and time delays in the optical fiber because different laser modes have slightly different wavelengths, which leads to fluctuations in the arrival time of the signal at the fiber output end [19]. As a result, the power fluctuations between laser modes produce an additional chromatic dispersion. When the power fluctuations among the dominant modes are quite large, significant variations in the signal power are detected at the receiver in optical links where the fiber has a large material dispersion. The power penalty due to the mode partition noise (in linear units) is described by

$$P_{\mathrm{MPN}} = \frac{1}{\sqrt{1-(Q\text{œ}_{\mathrm{MPN}})^2}}, \tag{7.3}$$

where Q is the value of the digital SNR and σ_{MPN} is the root mean squared (rms) value of the mode partition noise. The SNR due to the mode partition noise is independent of the signal power, and hence the BER of the total system cannot be improved without exceeding the limitation on the mode partition noise [20]. This is a significant difference from the degradation of receiver sensitivity, which can be compensated by increasing the signal power. Although the discussion here applies to edge-emitting lasers, it probably overestimates the penalty for VCSELs because the optical spectrum of the VCSEL is essentially limited. Thus, the VCSEL spectrum is not extended compared to that of an edge-emitting laser [21].

7.3.4 Relative Intensity Noise

Fluctuations in the output intensity of a laser result in optical intensity noise, which is usually called the *relative intensity noise* [22]. The relative intensity noise is usually expressed in decibels per hertz (dB/Hz), and all types of lasers generate this noise. The power penalty due to the relative intensity noise (in linear units) is described by

$$P_{\mathrm{RIN}} = \frac{1}{\sqrt{1-(Q\text{œ}_{\mathrm{RIN}})^2}}, \tag{7.4}$$

where Q is the value of the digital SNR, and σ_{RIN} is the rms value of the relative intensity noise.

7.4 Coupling Loss

Silica-based SMFs are widely used in long-haul networks because of their extremely low attenuation and high bandwidth [23]. However, silica-based SMF connections require accurate alignment using expensive and precise connectors, which increases the installation cost, especially in LANs where many connections are expected [24]. On the other hand, optical fibers are currently required for data communication in LANs in homes, offices, hospitals, vehicles, and aircraft, as well as in long-haul telecommunication. As optical signal processing and transmission speeds have increased with developments in information and communication technologies, metal wiring has become a bottleneck for high-speed data transmission systems and large parallel processing computer systems. This is because electrical wiring causes significant problems, including electromagnetic interference, high signal reflection, high power consumption, and heat generation [25]. GI POFs have been investigated as the transmission media in high-speed, short-reach networks [26]. This is because the GI POF has excellent characteristics, such as high bandwidth, a large core, high flexibility, and low bending loss, which allow rough connections and easy handling [27–32]. Therefore, the GI POF is expected to dramatically reduce the installation cost of networks, particularly LANs [33, 34]. Because a large core diameter is generally required for rough connections, the GI POF has an advantage for consumer use because of its large core.

A ball lens termination design using a ballpoint pen technique was proposed for the GI POF. The ballpoint pen technique demonstrated highly efficient coupling even though the connection was quite rough and a large core (e.g., several hundreds of micrometers) was not necessarily required for the rough connection. This connection with high coupling efficiency is called the *ballpoint pen interconnection.*

7.4.1 Core Diameter Dependence

The coupling loss of a fiber connection strongly depends on the core diameter of the optical fiber because the ratio of overlapping core regions is determined by the core diameter even for connections with the same misalignment. Figure 7.6 shows the coupling losses of three GI POFs with different core diameters under axial misalignment compared to that of an SI (step-index) POF. All the GI POFs were fabricated by changing the heat drawing ratio from the same preform prepared by the interfacial-gel polymerization (see Chapter 5). Therefore, the GI POFs had the same NA of 0.22 and the same RIP of $g = 5.7$, but different core diameters. The core diameters were estimated by the measured near-field pattern (NFP; see Chapter 6). The three GI POFs had core diameters of 240, 350, and 520 μm. The SI POF, commercially available from Mitsubishi Rayon, had a core diameter of 980 μm and an NA of 0.30. A remarkable reduction in the coupling loss was observed in the GI

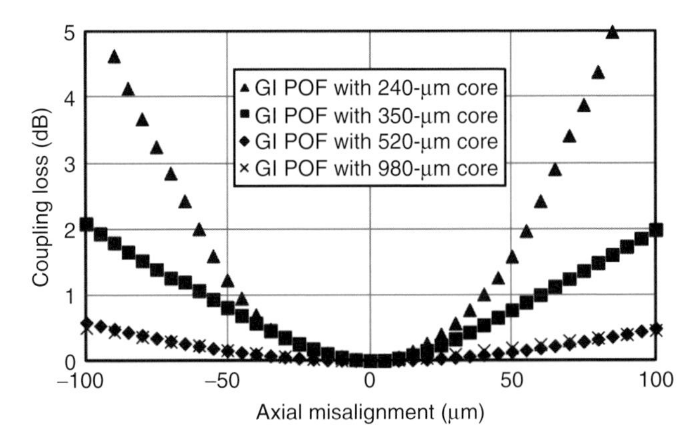

Figure 7.6 Coupling losses under axial misalignment of GI POFs with the same NA and RIP but different core diameters compared with that of SI POF.

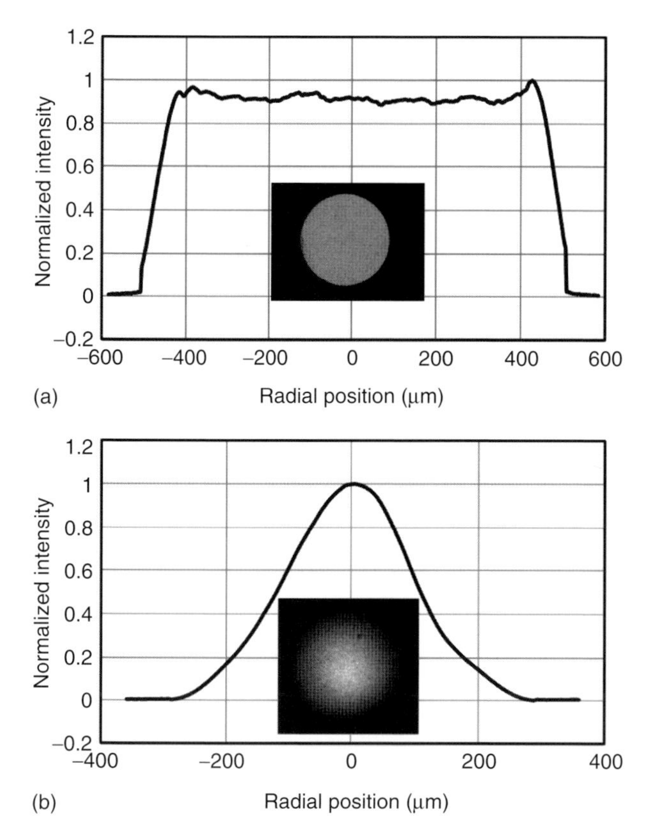

Figure 7.7 NFPs of (a) SI and (b) GI POFs. Inset shows a photograph of output beam from fiber end face.

POF as the core diameter was larger. On the other hand, the GI POF with a diameter of 520 μm exhibited almost the same coupling loss as the SI POF, although the core diameter of the GI POF was almost half that of the SI POF with a diameter of 980 μm. This is because the output optical power distribution of the SI POF is almost constant throughout the core region. However, most of the output power of the GI POF is confined to near the core center. Figure 7.7 shows the measured NFPs of the SI and GI POFs. A difference in the output optical power distributions is clearly observed. Therefore, the coupling efficiency of the GI POF is less sensitive to misalignment, and the GI POF exhibits a large tolerance for mechanical misalignment. This is a great advantage in the design of cost-effective, user-friendly optical networks.

7.4.2
Ballpoint Pen Termination

The advantages of a ball lens connection between optical fibers are schematically shown in Figure 7.8 in comparison with conventional direct butting. The output

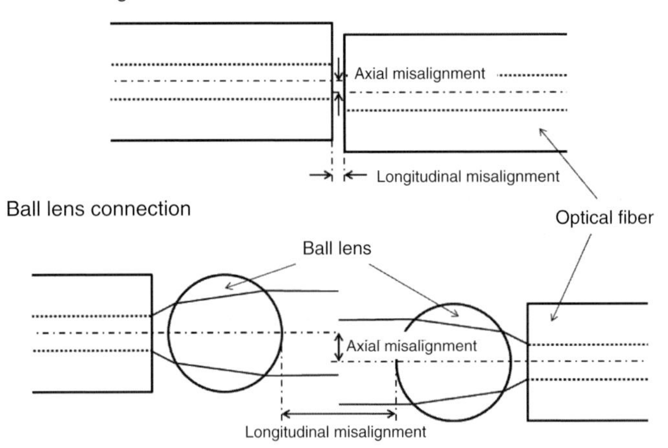

Figure 7.8 Concept of ball lens connection.

beam from the optical fiber is expanded and collimated by the ball lens, which dramatically increases the tolerance for axial and longitudinal misalignment of the optical axis [35]. The ball lens is not only a beam expander and collimator but also a shield of the fiber end face from scratches and dust. Although ball lens connections have these excellent characteristics [36], the termination process requires accurate alignment of the fiber and lens, significantly increasing the total cost of system installation [37].

A ball lens termination designed for GI POFs was proposed and fabricated by Keio University in collaboration with Mitsubishi Pencil, which was derived from a cost-effective and sophisticated technique proven in the mass production of ballpoint pens. The structure of the ballpoint pen termination for the GI POF is shown in Figure 7.9. A glass ball and an optical fiber are substituted for the metal ball and ink case of a ballpoint pen. The ball lens was precisely fabricated as a sphere by the ballpoint pen technique. This ballpoint pen technique also enables precise and cost-effective alignment of the optical axes of the GI POF and ball lens, although its structure is quite simple and compact. This is a great advantage in mounting and optical interconnection design. Here, the ball lens had a diameter of 550 μm and a refractive index of 1.51. A perfluorinated-polymer-based GI POF commercially available from Asahi Glass Co. was used. The GI POF had core and cladding diameters of 80 and 490 μm, respectively. The NA was 0.245. A data rate of 10 Gb/s for 100 m is certified in the catalog. The GI POF and ball lens were inserted into a metal sleeve and fixed by swaging the sleeve and by press-fitting, respectively.

The coupling losses of the GI POF with and without the ball lens were evaluated at a wavelength of 850 nm, as shown in Figure 7.10, under intentional axial and longitudinal misalignments. The coupling loss was dramatically decreased by the ballpoint pen termination. Without the ball lens, the coupling loss of the GI POF increased rapidly with increasing misalignment in both the axial and longitudinal cases. In contrast, the coupling loss with the ball lens was

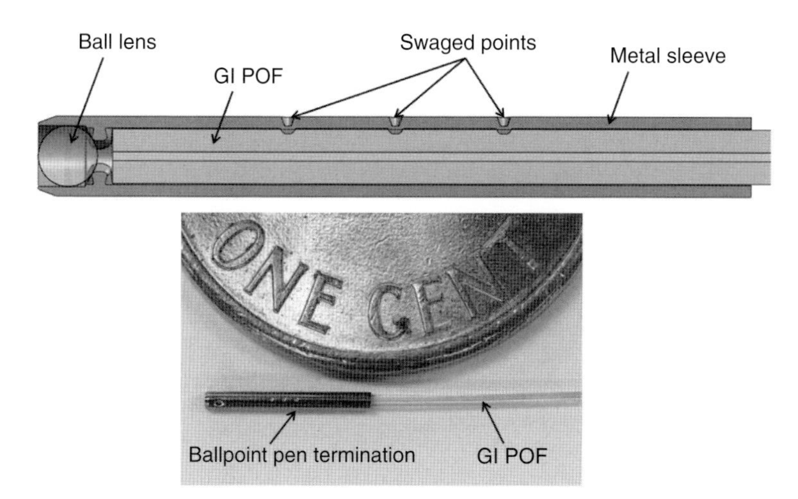

Figure 7.9 Structure of ballpoint pen termination.

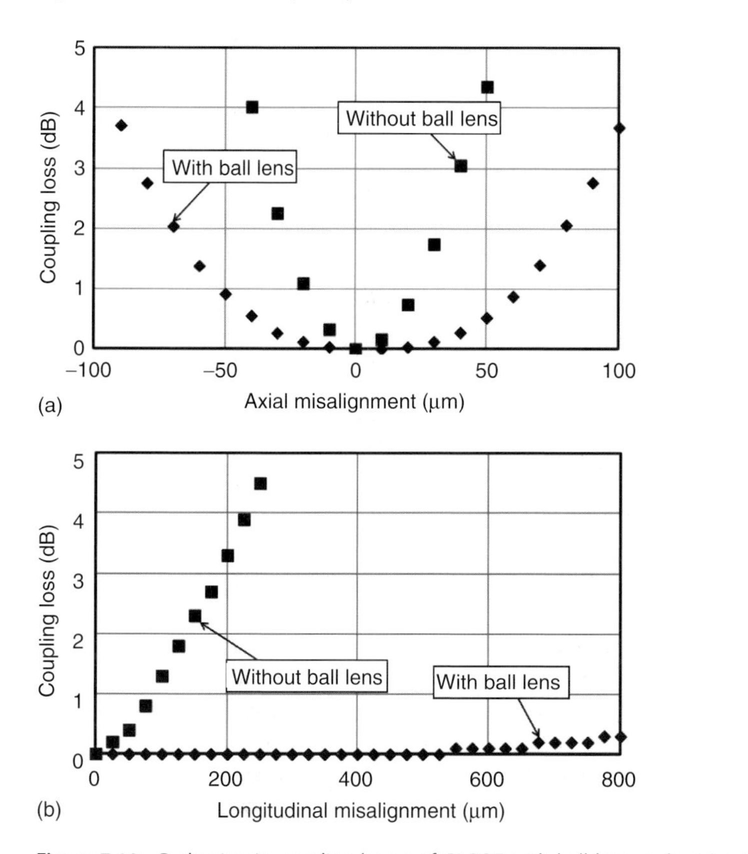

Figure 7.10 Reduction in coupling losses of GI POF with ball lens under (a) axial and (b) longitudinal misalignment.

less than 1.0 dB even when the axial misalignment was larger than 50 μm. No coupling loss was observed even when the longitudinal misalignment was over 500 μm. The ballpoint pen termination provided a much greater tolerance of axial misalignment and exhibited practically no coupling loss under longitudinal misalignment because the ball lens expanded and collimated the output beam from the GI POF. In addition, the ball lens protects the fiber end face from physical impacts and dust. Thus, the ballpoint pen termination realized rough connection, easy handling, and high coupling efficiency, and demonstrated that a large core (e.g., hundreds of micrometers) was not necessarily required for rough and cost-effective connection.

7.4.3 Ballpoint Pen Interconnection

A strong demand for greater realism in video images and face-to-face communication requires higher resolution, more natural color, and higher frame rates. Thus, higher and higher bit rates in data transmission are required in high-resolution displays and cameras.

A 120 Gb/s optical cable and interface are shown in Figure 7.11. The optical cable contained 12 GI POFs with ballpoint pen terminations and a 5-V power supply line. The connector size is a very important factor in the mounting design; the cross section of the compact connecter was 12.9 mm in width and 3.7 mm in thickness. Video data transmission in a format of 8K ultrahigh definition (8K UHD) was demonstrated using only one 120 Gb/s optical cable with the ballpoint pen interconnection, whereas 8K UHD video data transmission required 16 conventional high-definition serial digital interface (HD-SDI) cables. The optical cable with a compact interface was thin, lightweight, and flexible, and enabled rough connections. These characteristics are great advantages for consumer use.

Figure 7.11 120 Gb/s optical cable with ballpoint pen interconnection. 8K UHD video transmission was demonstrated with only one 120 Gb/s optical cable. 8K video transmission required 16 conventional HD-SDI cables.

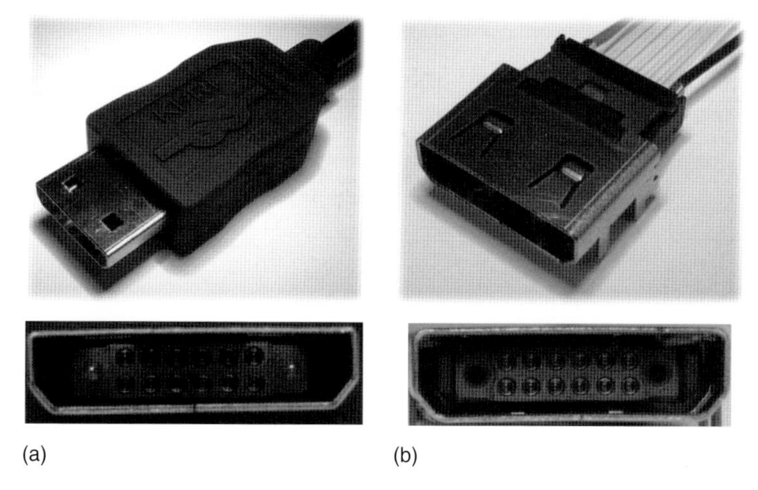

(a) (b)

Figure 7.12 Plug (a) and receptacle (b) with 12 channels of high-speed GI POFs and ballpoint pen interconnections.

Photographs of the plug and receptacle with high-speed GI POF and ballpoint pen interconnections are shown in Figure 7.12. The connection loss of the connector was evaluated in practical use, not under ideal conditions, such as the use of a V-groove mounted on three-dimensional and angular micropositioners. The setup for measuring the connection loss of the connector is shown in Figure 7.13. In this measurement, an optical beam from a light-emitting diode with an emission wavelength of 850 nm was first injected into a 2-m silica-based MMF with a core diameter of 50 μm, and then the output beam from the silica-based MMF was coupled to the perfluorinated-polymer-based GI POF of the plug. The output beam from the GI POF of the plug was coupled to the GI POF of the receptacle through the ballpoint pen interconnection. The connection loss of this connector was evaluated in terms of the difference in the optical powers at two measurement points, P_1 and P_2. P_1 is the output optical power from the silica-based MMF, and P_2 is that from the GI POF of the receptacle. Thus, the measured connection loss included not only the coupling loss between the plug and the receptacle with the ballpoint pen interconnection but also the insertion losses of the ballpoint pen terminations, the coupling loss from the MMF to the GI POF of the plug, and the transmission loss of the GI POF.

The connection losses of 15 pairs of plugs and receptacles for optical cables with ballpoint pen interconnections were evaluated. The variation in the connection losses among the sample connectors was negligible. No significant dependence of the connection losses on the channel positions in the connector was observed, as shown in Figure 7.14. Most of the connection losses were less than 1.0 dB, and the average loss was 0.88 dB. Thus, the ballpoint pen interconnection realized high coupling efficiency despite rough connections even in a practical application such as a thin, compact connector.

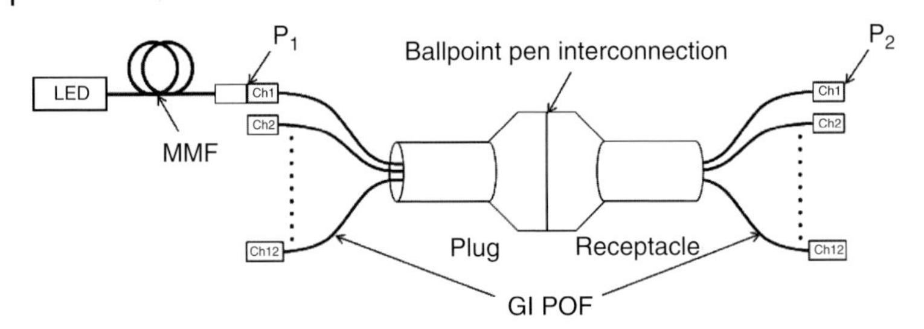

Figure 7.13 Setup for measuring connection loss of ballpoint pen interconnection.

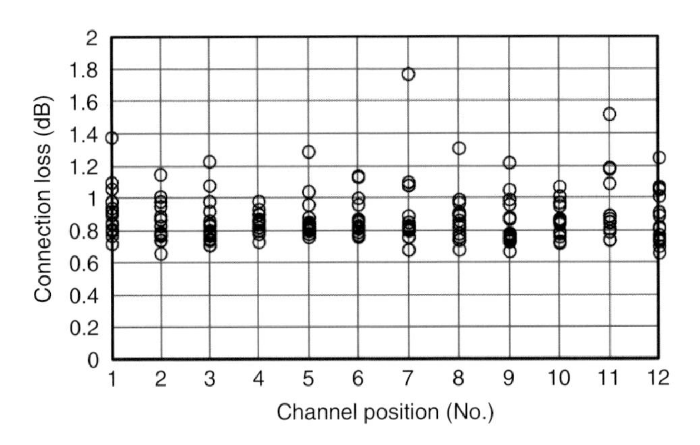

Figure 7.14 Connection losses versus channel position of 15 optical cables with ballpoint pen interconnections.

7.5
Design for Gigabit Ethernet

Gigabit Ethernet allows a power budget of 7.5 dB for MMF networks [5]. In the MMF link, ISI can contribute a large power penalty, as mentioned above. The maximum acceptable penalty due to ISI for Gigabit Ethernet is only 3.6 dB. Figure 7.15 shows the estimated power penalty of the GI POF as a function of the RIP based on the link power budget model discussed in the preceding sections compared to the bandwidth calculated from the power-law RIP. The power penalty was calculated at a BER of 10^{-12}, the maximum allowed by Gigabit Ethernet. The GI POF was assumed to have an NA of 0.20 and a length of 100 m. Note that the link power penalty is strongly influenced by the transmitter and receiver; this estimation is just an example. A power penalty of less than 3.6 dB indicates that a bandwidth higher than 600 MHz is acceptable, which corresponds to a range of the profile exponent g from 1.7 to 3.5 when the fiber length is 100 m. Therefore, the GI POF can provide a 100-m power-penalty-free Gigabit Ethernet link, and hence large tolerances of the other components in the link design, for example, the use of an inexpensive and inaccurate connector.

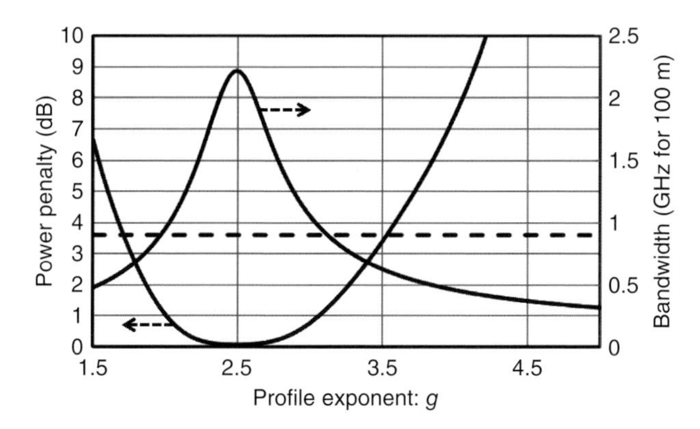

Figure 7.15 Design of GI POF acceptable for Gigabit Ethernet based on link power budget. Broken line denotes limit on power penalty due to the ISI specified by Gigabit Ethernet.

Although the narrow bandwidth of the SI POF is unacceptable for Gigabit Ethernet, there are many reports of Gigabit Ethernet link designs consisting of an SI POF with an equalizer, FEC, and orthogonal frequency-division multiplexing (OFDM) [38, 39]. The equalizer can reshape the distorted pulses as they travel through the fiber and can decrease the effects of ISI. Equalizers can be implemented to increase the bandwidth dramatically. Thus, the equalizer compensates for the poor bandwidth of the SI POF and improves the BER.

References

1. Kimura, T. (1988) Factors affecting fiber-optic transmission quality. *J. Lightwave Technol.*, **6** (5), 611–619.
2. Lowery, A., Lenzmann, O., Koltchanov, I., Moosburger, R., Freund, R., Richter, A., Georgi, S., Breuer, B., and Hamster, H. (2000) Multiple signal representation simulation of photonic devices, systems, and networks. *IEEE J. Sel. Top. Quantum Electron.*, **6** (2), 282–296.
3. Nowell, M.C., Cunningham, D.G., Hanson, D.C., and Kazovsky, L.G. (2000) Evaluation of Gb/s laser based fibre LAN links: review of the Gigabit Ethernet model. *Opt. Quantum Electron.*, **32** (2), 169–192.
4. Kim, D., Kim, H., and Eo, Y. (2012) Analytical eye-diagram determination for the efficient and accurate signal integrity verification of single interconnect lines. *IEEE Trans. Comput. Aided Des. Integr. Circuits Syst.*, **31** (10), 1536–1545.
5. Cunningham, D. and Lane, W. (1999) *Gigabit Ethernet Networking*, Macmillan Publishing Co., Inc.
6. Yamamoto, Y. (1980) Receiver performance evaluation of various digital optical modulation-demodulation systems in the 0.5–10 μm wavelength region. *IEEE J. Quantum Electron.*, **16** (11), 1251–1259.
7. Agrawal, G.P., Anthony, P.J., and Shen, T.M. (1988) Dispersion penalty for 1.3 μm lightwave systems with multimode semiconductor lasers. *J. Lightwave Technol.*, **6** (5), 620–625.
8. Rodes, R., Mueller, M., Bomin, L., Estaran, J.B., Jensen, J., Gruendl, T., Ortsiefer, M., Neumeyr, C., Rosskopf, J., Larsen, K.J., Amann, M.C., and Monroy, I.T. (2013) High-speed 1550 nm VCSEL data transmission link employing 25 Gbd 4-PAM modulation and hard decision forward error correction. *J. Lightwave Technol.*, **31** (4), 689–695.

9. Sakaguchi, J., Awaji, Y., Wada, N., Kanno, A., Kawanishi, T., Hayashi, T., Taru, T., Kobayashi, T., and Watanabe, M. (2011) 109-Tb/s (7× 97× 172-Gb/s SDM/WDM/PDM) QPSK transmission through 16.8-km homogeneous multi-core fiber. Proceedings of the OFC/NFOEC.
10. Bosco, G., Carena, A., Curri, V., Gaudino, R., and Poggiolini, P. (2003) Quantum limit of direct-detection receivers using duobinary transmission. *IEEE Photonics Technol. Lett.*, **15** (1), 102–104.
11. Tayahi, M.B., Fan, H., Webster, R., and Dutta, N.K. (1998) Digital and analog transmission through polymer optical fiber. *Proc. SPIE*, **3414**, 139–148.
12. Kim, H. and Gnauck, A.H. (2003) Experimental investigation of the performance limitation of DPSK systems due to nonlinear phase noise. *IEEE Photonics Technol. Lett.*, **15** (2), 320–322.
13. Wan, P. and Conradi, J. (1996) Impact of double Rayleigh backscatter noise on digital and analog fiber systems. *J. Lightwave Technol.*, **14** (3), 288–297.
14. Personic, S.D. (1973) Receiver design for digital fiber optic communication systems I. *Bell Syst. Tech. J.*, **52** (6), 843–874.
15. Personic, S.D. (1973) Receiver design for digital fiber optic communication systems II. *Bell Syst. Tech. J.*, **52** (6), 875–886.
16. Liu, F., Rasmussen, C.J., and Pedersen, R.J.S. (1999) Experimental verification of a new model describing the influence of incomplete signal extinction ratio on the sensitivity degradation due to multiple interferometric crosstalk. *IEEE Photonics Technol. Lett.*, **11** (1), 137–139.
17. Ogawa, K. and Vodhanel, R.S. (1982) Measurements of mode partition noise of laser diodes. *IEEE J. Quantum Electron.*, **18** (7), 1090–1093.
18. Jensen, N., Olesen, H., and Stubkjaer, K.E. (1987) Partition noise in semiconductor lasers under CW and pulsed operation. *IEEE J. Quantum Electron.*, **23** (1), 71–80.
19. DiDomenico, M. (1972) Material dispersion in optical fiber waveguides. *Appl. Opt.*, **11** (3), 652–654.
20. Olsson, N.A., Tsang, W.T., Temkin, H., Dutta, N.K., and Logan, R.A. (1985) Bit-error-rate saturation due to mode-partition noise induced by optical feedback in 1.5-μm single longitudinal-mode C^3 and DFB semiconductor lasers. *J. Lightwave Technol.*, **3** (2), 215–218.
21. Law, J.Y. and Agrawal, G.P. (1997) Mode-partition noise in vertical-cavity surface-emitting lasers. *IEEE Photonics Technol. Lett.*, **9** (4), 437–439.
22. Kallimani, K.I. and O'Mahony, M. (1998) Relative intensity noise for laser diodes with arbitrary amounts of optical feedback. *IEEE J. Quantum Electron.*, **34** (8), 1438–1446.
23. Olshansky, R. (1979) Propagation in glass optical waveguides. *Rev. Mod. Phys.*, **51** (2), 341–367.
24. Polishuk, P. (2006) Plastic optical fibers branch out. *IEEE Commun. Mag.*, **44** (9), 140–148.
25. Ball, P. (2012) Computer engineering: feeling the heat. *Nature*, **492** (7428), 174–176.
26. Koike, Y. (1991) High-bandwidth graded-index polymer optical fibre. *Polymer*, **32** (10), 1737–1745.
27. Zubia, J. and Arrue, J. (2001) Plastic optical fibers: an introduction to their technological processes and applications. *Opt. Fiber Technol.*, **7** (2), 101–140.
28. Ishigure, T., Hirai, M., Sato, M., and Koike, Y. (2004) Graded-index plastic optical fiber with high mechanical properties enabling easy network installations. *J. Appl. Polym. Sci.*, **91** (1), 404–416.
29. Makino, K., Ishigure, T., and Koike, Y. (2006) Waveguide parameter design of graded-index plastic optical fibers for bending-loss reduction. *J. Lightwave Technol.*, **24** (5), 2108–2114.
30. Makino, K., Kado, T., Inoue, A., and Koike, Y. (2012) Low loss graded index polymer optical fiber with high stability under damp heat conditions. *Opt. Express*, **20** (12), 12893–12898.
31. Makino, K., Akimoto, Y., Koike, K., Kondo, A., Inoue, A., and Koike, Y. (2013) Low loss and high bandwidth polystyrene-based graded index polymer optical fiber. *J. Lightwave Technol.*, **31** (14), 2407.

32. Makino, K., Nakamura, T., Ishigure, T., and Koike, Y. (2005) Analysis of graded-index polymer optical fiber link performance under fiber bending. *J. Lightwave Technol.*, **23** (6), 2062–2072.

33. Koike, Y., Ishigure, T., and Nihei, E. (1995) High-bandwidth graded-index polymer optical fiber. *J. Lightwave Technol.*, **13** (7), 1475–1489.

34. Koike, Y. and Ishigure, T. (2006) High-bandwidth plastic optical fiber for fiber to the display. *J. Lightwave Technol.*, **24** (12), 4541–4553.

35. Baker, J.C. and Payne, D.N. (1981) Expanded-beam connector design study. *Appl. Opt.*, **20** (16), 2861–2867.

36. Drake, M.D. (1985) A critical review of fiber optic connectors. Proceedings of the 28th Annual Technical Symposium, pp. 57–69.

37. Chandrappan, J., Jing, Z., Mohan, R.V., Gomez, P.O., Aung, T.A., Yongfei, X., Ramana, P.V., Lau, J.H., and Kwong, D.L. (2009) Optical coupling methods for cost-effective polymer optical fiber communication. *IEEE Trans. Compon. Packag. Technol.*, **32** (3), 593–599.

38. Atef, M., Swoboda, R., and Zimmermann, H. (2012) 1.25 Gbit/s over 50 m step-index plastic optical fiber using a fully integrated optical receiver with an integrated equalizer. *J. Lightwave Technol.*, **30** (1), 118–122.

39. Wei, J.L., Geng, L., Cunningham, D.G., Penty, R.V., and White, I.H. (2012) Comparisons between gigabit NRZ, CAP and optical OFDM systems over FEC enhanced POF links using LEDs. Proceedings of the 14th International Conference on Transparent Optical Networks (ICTON).

Appendix
Progress in Low-Loss and High-Bandwidth Plastic Optical Fibers*

A.1
Introduction

One of the Nobel Prizes in physics in 2009 was awarded to Charles Kao for "groundbreaking achievements concerning light transmission in fibers for optical communication." In the 1960s, glass optical fibers (GOFs) that could be used for transmitting light had already been developed [2, 3]. However, these fibers exhibited significant propagation losses above 1000 dB/km; thus the transmission distance was severely limited. Kao made a discovery that led to a breakthrough in fiber optics. After closely examining the possibilities for reducing propagation losses, he realized that the attenuation of existing fibers were orders of magnitude above the fundamental limit [4]. He theorized that by using a fiber of the purest glass, it would be possible to transmit light signals over 100 km, in contrast to fibers available in the 1960s that could only realize a distance of 20 m. His prediction of the future potential of low-loss optical fibers inspired other researchers to make great efforts toward its realization. Just 4 years later, in 1970, the first ultrapure fiber with an attenuation of 20 dB/km was successfully fabricated [5]. Others soon followed, and losses were pushed down to the theoretical limit (0.2 dB/km) [6–8]. A first-generation fiber-optic communication system was successfully deployed in 1976, and since then GOFs have steadily connected the world. Today, over a billion kilometers of GOFs, sufficient to encircle the globe more than 25 000 times, have been placed underground and in oceans, enabling global broadband communication media such as the Internet.

We believe that the next big step in optical fiber technology will be plastic optical fibers (POFs). In advanced countries, "fiber-to-the-home" (FTTH) services that interconnect homes with a GOF backbone are well established. However, intra-building networks such as home networks have yet to be adequately developed. With the increase in the number and use of home PCs, high-definition (3D) TVs, blu-ray devices, digital cameras, storage devices, and intelligent home appliances, the demand has increased for optical data-link connections not only up to the

* This review was previously published in the Journal of Polymer Science: Part B: Polymer Physics [1].

Fundamentals of Plastic Optical Fibers, First Edition. Yasuhiro Koike.

end-users' buildings but also within them [9]. If we compare communication networks to human blood vessels, the GOF backbone functions like the arteries and veins, whereas the home network is similar to the capillaries. Surprisingly, the entire length of all terminal networks accounts for 95% of all optical networks. In such short-distance applications, the extremely low attenuation and enormous capacity of a single-mode GOF is unnecessary. Instead, simpler and less expensive components, greater flexibility, and higher reliability against bending, shocks, and vibrations are considerably more valuable properties. Naturally, polymers are overwhelmingly superior to glasses in all these requirements.

However, for POFs, which have been overshadowed by the success of GOFs in previous decades, the road has never been smooth. In particular, many studies have sought to reduce fiber attenuation as had been done for GOFs. To be used as the core base material, polymers must ultimately be transparent. With its amorphous structure and easy processing via a free-radical bulk polymerization that requires no metal catalysts and avoids contaminants, poly(methyl methacrylate) (PMMA) has mainly been utilized for POFs [10]. Coupled with low material costs, its adequate thermal stability and excellent corrosion resistance have also made PMMA the representative material for POFs. However, even with highly purified PMMA-based POFs, the intrinsic problem of attenuation caused by C–H molecular vibrations remained unresolved. Furthermore, POFs with a large core and modal dispersion had severe limitations in bandwidth. In this review, we begin with a brief explanation of optical fibers, and proceed to describe technical developments that have solved the above-mentioned problems and propelled POFs into the main stream of modern home networking applications.

A.2
Basic Concept and Classification of Optical Fibers

The principle of light propagation through POFs is the same as that for GOFs. POFs are basically composed of two coaxial layers: the core and the cladding. The core is the inner part of the fiber that guides light, whereas the cladding completely surrounds the core. The refractive index of the core is slightly higher than that of the cladding. Hence, when the incident angle of the light input to the core is greater than the critical angle, the input is confined to the core region because of total internal reflection at the interface between the different dielectric materials comprising each layer. Although this simple concept is a useful approximation of light guidance in many kinds of fibers, it does not provide a full explanation. Light is really an electromagnetic wave with a frequency in the optical range. An optical fiber guides the waves in distinct patterns called *modes*, which describe the distribution of light energy across the waveguide. Commonly used optical fibers can be separated into two classes based on their modal properties: single-mode and multimode fibers. Single-mode fibers are step-index (SI) fibers, whereas multimode fibers can be divided into SI and graded-index (GI) fibers. SI and GI refer to patterns of variation in the refractive index with the radial distance from the

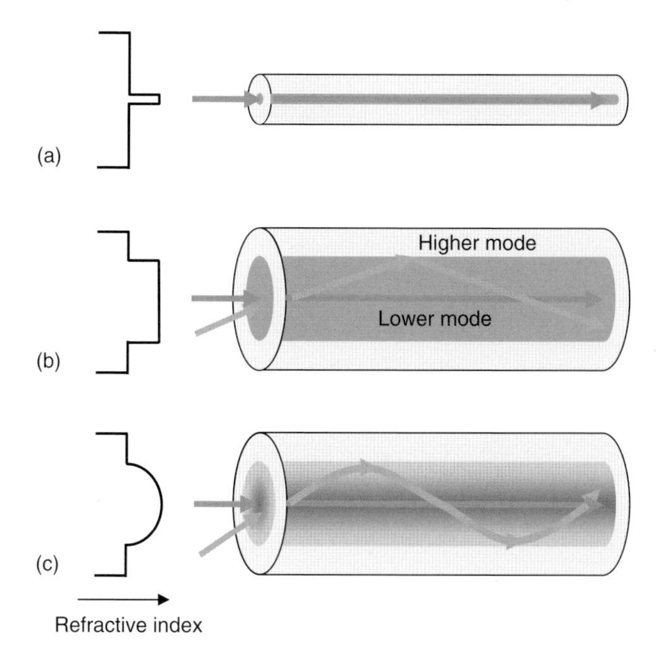

Figure A.1 Ray trajectories through basic types of optical fibers. (a) Single-mode fiber. (b) Step-index multimode fiber. (c) Graded-index multimode fiber.

fiber axis. Figure A.1 schematically shows these three types of fibers: (a) the single-mode fiber, (b) the SI multimode fiber, and (c) the GI multimode fiber.

The information-carrying capacity of an optical fiber is determined by its impulse response. The impulse response and hence the bandwidth are largely determined by the modal properties of the fiber. Because single-mode transmission avoids modal dispersion and other effects that occur with multimode transmission, single-mode fibers with core diameters of 5–10 μm can carry signals at considerably higher speeds than multimode fibers [11, 12]. Modal dispersion can be understood by referring to the SI multimode fiber (Figure A.1b), where different rays are shown to travel along paths with different lengths. Even though they are coincident at the input end and travel at the same speed within the fiber, these rays disperse at the output end because of their different path lengths. As a result, the impulse signal broadens. This becomes a serious restriction on transmission speed because pulses that overlap can interfere with each other, making it impossible to receive the signal. On the other hand, GI guides have less modal dispersion and greater transmission capacity than SI guides [13–16]. The refractive index of the core in GI multimode fibers is not constant but decreases gradually from its maximum at the core center to its minimum at the core/cladding boundary. From Figure A.1c, it is easy to understand qualitatively why the modal dispersion decreases for GI fibers. As in the case of SI fibers, the path is longer for more oblique rays (higher order modes). However, the ray velocity changes along the path because of variations in the refractive index. The speed of light in a material, v, is equal to the velocity of light in vacuum, c, divided by the refractive index, that is, $v = c/n$. More specifically, ray propagation along the fiber axis takes the shortest path but has the slowest speed since the index is largest along this path. Oblique rays have a large part of their path in a lower refractive index medium in which they travel faster. The difference in the refractive index is small but sufficient to compensate for the time delay. By carefully controlling the refractive index profile in the core region, the modal dispersion can be drastically reduced.

Currently, most GOFs are fabricated with single-mode structures and are commonly used in long-distance applications such as core and metropolitan networks. With respect to coupling loss, the small core of single-mode fibers is a serious disadvantage; the smaller the core diameter, the harder it is to couple light into the fiber. Hence, GI multimode-type fibers were also extensively studied [13–16] and deployed in some telecommunication applications up until the mid-1980s. However, as single-mode GOFs were far superior in both attenuation and bandwidth, they gradually shifted to short-length applications such as storage area networks. Coupling light into a single-mode fiber inevitably requires considerably tighter tolerances than doing so into the larger cores of a multimode fiber. However, such tighter tolerances were achieved; nowadays, the single-mode GOF has become the standard choice for virtually all kinds of telecommunications that involve high bit rates or span distances longer than a couple of kilometers.

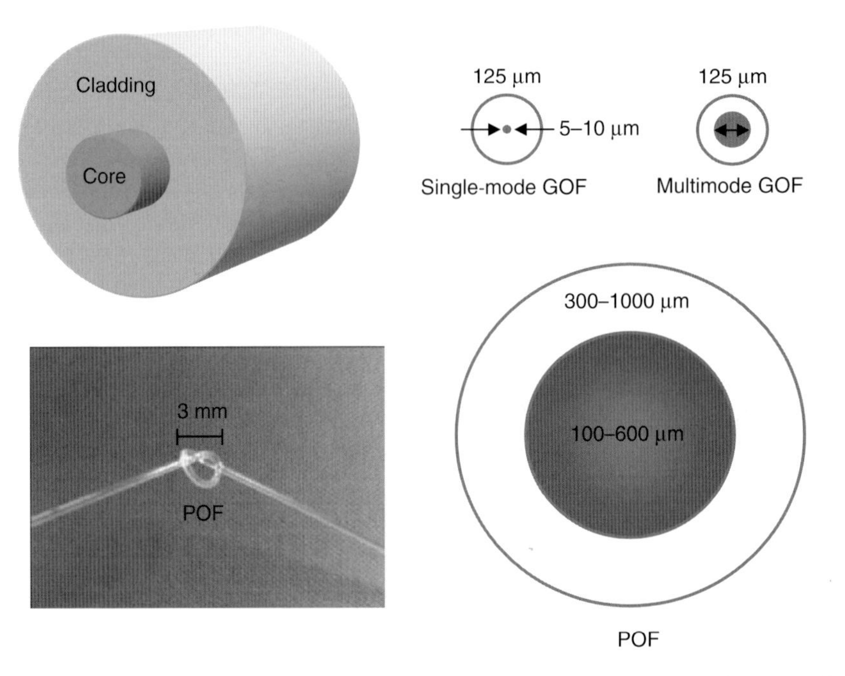

Figure A.2 Cross-sectional views of representative optical fibers and a photo of a POF with a 3-mm knot.

On the other hand, POFs have attracted attention as optimal candidates for short-distance networks such as intra-building networks [10]. Optical fibers must be able to connect various devices in the individual rooms of a building; hence, they must be flexible, easy to bend, and connected at several points. Predictably, the single-mode GOF is unsuited for this application given its brittleness and small core. In contrast, POFs can be enlarged to ~1 mm in diameter without losing flexibility or ease of fiber alignment (Figure A.2). Therefore, apart from a few exceptions [17–21], a great deal of effort has been focused on developing multimode POFs with large diameters.

A.3
The Advent of Plastic Optical Fibers and Analysis of Attenuation

The first POF, Crofon™, which was invented in the mid-1960s by Du Pont, was a multimode fiber with an SI profile in the core region [22, 23]. Compared to GOFs, the SI POF was also advantageous in terms of mass production; it was not only inexpensive to fabricate but also easy to mold and manufacture. The first commercialized SI POF was Eska™, which was introduced by Mitsubishi Rayon in 1975; [24, 25] subsequently, Asahi Chemical and Toray also entered the market. However, the first fibers were not sufficiently transparent to be used as an intra-building communication medium, and their application was severely limited to extremely short-range areas such as light guides, illuminations, audio data-links, and sensors.

Fiber attenuation limits how far a signal can propagate in the fiber before the optical power becomes too weak to be detected. The oldest and most common approach for determining fiber attenuation is measuring the optical power transmitted through long and short lengths of the same fiber using identical input couplings [26], a method known as the *cutback technique*. First, the optical power at the output (or far end) of the fiber is measured. Subsequently, the fiber is cut off a few meters from the light source, and the output power at this near end is measured without disturbing the input. When P_{out} and P_{in} represent the output powers of the far and near ends of the fiber, respectively, the attenuation α, which is normally expressed in decibels, is defined as follows:

$$\alpha\ (\mathrm{dB/km}) = -\frac{10}{L}\log_{10}\left(\frac{P_{out}}{P_{in}}\right), \tag{A1}$$

where L (km) is the distance between the two measurement points. To estimate the acceptable transmission losses of the fiber, the total optical link from the light source to the receiver must be considered. As a typical example for home networking, the minimum output power of the light source is −6 dBm and the lowest optical sensitivity of the receiver is −18 dBm. Considering other factors such as coupling loss (1.5 dB), bending loss (0.5 dB), power penalty (1.5 dB), and margin (2.5 dB), the acceptable transmission loss for the fiber is 6 dB. Accordingly, in the case of deploying POFs in a home, in which the maximum laying distance is ~30 m,

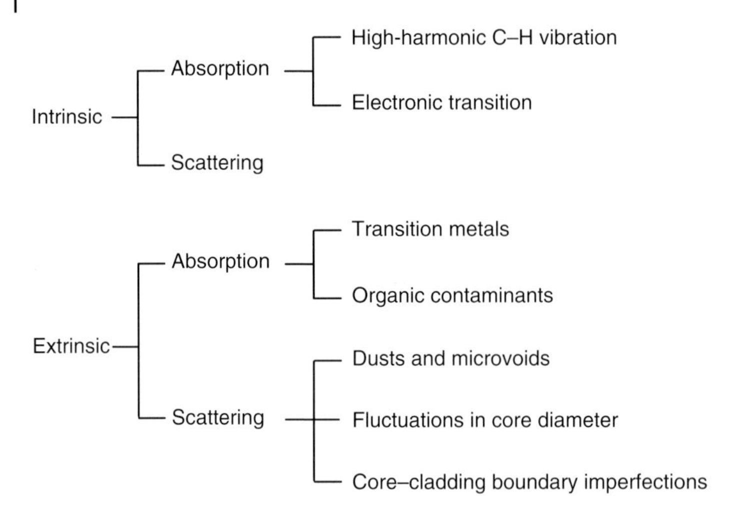

Figure A.3 Classification of intrinsic and extrinsic factors affecting POF attenuation.

the fiber attenuation must be <200 dB/km. In comparison, the attenuation of the first prototype was over 1000 dB/km. Faced with this challenge, many researchers have attempted to reduce the attenuation. Although the various mechanisms contributing to losses in POFs are basically similar to those in GOFs, their relative magnitudes differ. Figure A.3 shows the loss factors for POFs, which are divided into intrinsic and extrinsic factors. These are further classified into absorption and scattering losses.

Through analyses of each factor of fiber attenuation described below, the limitations of POFs based on several materials and their theoretical groundings have been steadily clarified. Meanwhile, it has been shown that the major factors causing high attenuation were not intrinsic but rather extrinsic – those such as contaminants becoming mixed into the polymer during the fiber fabrication process. By preparing the fiber in an all-closed system from monomer distillation to fiber drawing, low-loss POFs with losses near theoretical limits were fabricated. In 1981, Kaino reported a poly(styrene) (PSt)-based SI POF with an attenuation of 114 dB/km at 670 nm [27], and in the following year his group also succeeded in obtaining a PMMA-based SI POF with an attenuation of 55 dB/km at 568 nm [28]. Table A.1 is a brief early chronology of SI POFs with reduced attenuation.

A.3.1 Absorption Loss

Every material absorbs light energy, whose amount depends on the wavelength and the material. Intrinsic absorption loss in POFs is caused by electronic transitions and molecular vibrations. Electronic transition absorption results from transitions between electronic energy levels of bonds within the materials. The absorption of photons causes an upward transition, which leads to excitation of

Table A.1 Early history of loss reduction for SI POFs.

Year	Company	Core material	Attenuation	Wavelength (nm)
1968	DuPont	PMMA	1000	650
1972	Toray	PSt	1100	670
1978	Mitsubishi Rayon	PMMA	300	650
1981	NTT	PSt	114	670
1981	NTT	PMMA	55	568
1983	Mitsubishi Rayon	PMMA	110	570
1985	Asahi Chemical	PMMA	80	570
1986	Fujitsu	PC	450	770

the electronic state. Typically, electronic transition peaks appear in the ultraviolet wavelength region, and their absorption tails influence the transmission losses in POFs. For example, when azo compounds are used as the initiator, a PMMA core POF, one of the commercially available POFs, has the $n-\pi^*$ transition of the ester groups in MMA (methyl methacrylate) molecules, the $n-\sigma^*$ transition of the S–H bonds in chain-transfer agents, and the $\pi-\pi^*$ transition of the azo groups. The most significant absorption is the transition of the $n-\pi^*$ orbital of the double bond within the ester group. The relationship between the electronic transition loss α_e (dB/km) and the wavelength of the incident light λ (nm) can be expressed by Urbach's rule [29]:

$$\alpha_e = A\exp\left(\frac{B}{\lambda}\right). \tag{A2}$$

Here, A and B are substance-specific constants. In the case of PMMA, A and B have been clarified to be 1.58×10^{-12} and 1.15×10^4, respectively [30]. Hence, the α_e value of PMMA is less than 1 dB/km at 500 nm. On the other hand, PSt andpoly(carbonate) (PC), which are also universal polymers for POFs, exhibit considerably larger absorption losses (Figure A.4) [30, 31]. This is because of the $\pi-\pi^*$ transition of the phenyl groups in PSt and PC. The energy bandgap between the π and π^* levels is sufficiently small to be excited by visible light. The tails are drastically shifted to longer wavelengths as the conjugation length increases.

The effect of molecular vibrational absorptions becomes strong at wavelengths in the visible to the infrared region. In the case of PMMA, which shows negligibly small electronic transition absorption, this molecular vibration is the predominant factor contributing to fiber attenuation. The energy level of the absorption wavelength, which reflects the light absorption at each wavelength, is expressed by [32, 33]

$$G(\upsilon) = \nu_0\left(\upsilon+\frac{1}{2}\right) - \nu_0\chi\left(\upsilon+\frac{1}{2}\right)^2. \tag{A3}$$

Here, υ is the quantum number (= 0, 1, 2, 3 ...), χ is the anharmonic constant, and ν_0 is the fundamental frequency. The first and second terms on the right-hand side are the harmonic and anharmonic vibrations, respectively. The overtone

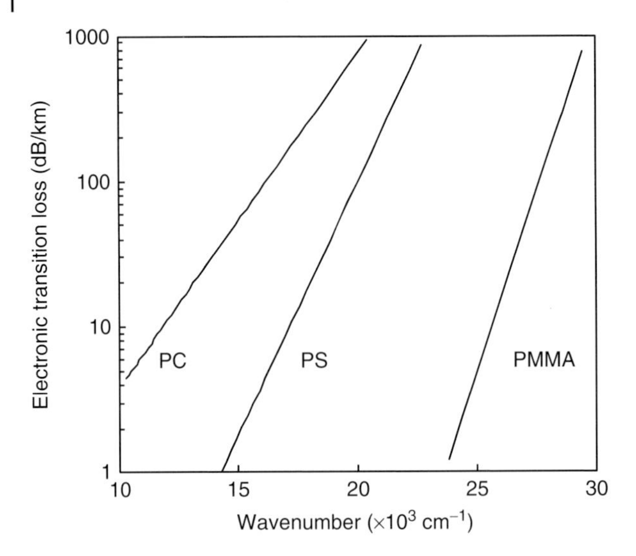

Figure A.4 Electronic transition losses for PMMA, PSt, and PC.

frequency is expressed as follows:

$$\nu_{\upsilon} = G(\upsilon) - G(0) = \nu_0 \upsilon - \chi \nu_0 \upsilon(\upsilon + 1). \tag{A4}$$

Setting the quantum number as 1, υ_0 is transformed as

$$\nu_0 = \frac{\nu_1}{1 - 2\chi}. \tag{A5}$$

Equation A5 can then be substituted into Equation A4 to yield

$$\nu_{\upsilon} = \frac{\nu_1 \upsilon - \nu_1 \chi \upsilon(\upsilon + 1)}{1 - 2\chi} \quad (\upsilon = 2, 3, 4 \ldots). \tag{A6}$$

The overtone absorption frequencies can be calculated using Equation A6 and χ, which can be obtained from measurements of ν_1 and the first overtone ν_2. Groh *et al.* estimated the attenuation by the overtone vibrational absorptions of the C–H and C–X bonds as follows [33]:

$$\alpha_{\mathrm{v}} = 3.2 \times 10^8 \frac{\rho N_{\mathrm{CX}}}{M} \left(\frac{E_{\upsilon}^{\mathrm{CX}}}{E_1^{\mathrm{CH}}} \right). \tag{A7}$$

Here, α_{v} (dB/km) is the attenuation, ρ (g/cm^3) is the polymer density, M (g/mol) is the molecular weight of the monomer unit, N_{CX} is the number of C–X bonds per monomer, and $E_{\upsilon}^{\mathrm{CX}}/E_1^{\mathrm{CH}}$ is the vibration energy ratio of each bond to the fundamental frequency of the C–H bond. Figure A.5 shows the spectral overtone positions and normalized integral band strengths for the C–H, C–D, and C–F vibrations. If we set $\rho = 1.19\,\mathrm{g/cm^3}$, $M = 100\,\mathrm{g/mol}$, and $N_{\mathrm{CH}} = 8$ as the PMMA values, it follows from Equation A7 that $E_{\upsilon}^{\mathrm{CH}}/E_1^{\mathrm{CH}} = 3.3 \times 10^{-8}$, which corresponds to an attenuation of 1 dB/km. In the visible to the near-infrared region, the overtones for C–D and C–F are several orders of magnitude lower than the overtone

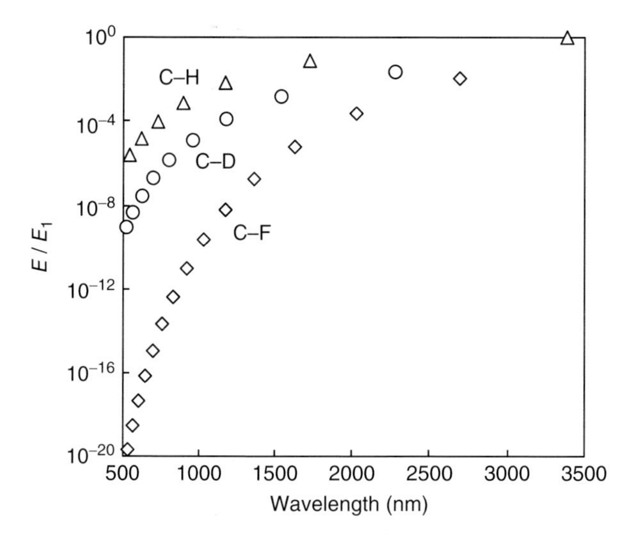

Figure A.5 Calculated spectral overtone positions and normalized integral band strengths for different C–X vibrations.

for C–H. This implies that fiber attenuation can be reduced drastically by substituting the hydrogen atoms by heavier atoms such as deuterium and fluorine with lower energy absorption bands.

A.3.2
Scattering Loss

Scattering losses in polymers arise from microscopic variations in material density. When natural light of intensity I_0 passes through a distance y, and its intensity is reduced to I by scattering loss, the turbidity τ is defined by

$$\frac{I}{I_0} = \exp(-\tau y). \tag{A8}$$

Since τ corresponds to the summation of all light scattered in all directions, it is given as

$$\tau = \pi \int_0^{\pi} (V_{\mathrm{V}} + V_{\mathrm{H}} + H_{\mathrm{V}} + H_{\mathrm{H}}) \sin\theta \, d\theta. \tag{A9}$$

Here, V and H denote vertical and horizontal polarizations, respectively. The symbol A and the subscript B in the expression for a scattering component A_B represent the directions of the polarizing phase of a scattered light and an incident light, respectively. θ is the scattering angle with respect to the direction of the incident ray. In structureless liquids or randomly oriented bulk polymers, these intensities are given by the following equations:

$$H_{\mathrm{V}} = V_{\mathrm{H}}, \tag{A10}$$

$$H_{\mathrm{H}} = V_{\mathrm{V}}\cos^2\theta + H_{\mathrm{V}}\sin^2\theta. \tag{A11}$$

Here, the isotropic part $V_{\mathrm{V}}^{\mathrm{iso}}$ of V_{V} is given as [34]

$$V_{\mathrm{V}}^{\mathrm{iso}} = V_{\mathrm{V}} - \frac{4}{3}H_{\mathrm{V}}. \tag{A12}$$

By substituting Equations A10–A12 into Equation A9, τ can be rewritten as follows:

$$\tau = \pi \int_0^{\pi} \left\{ \left(1 + \cos^2\theta\right) V_{\mathrm{V}}^{\mathrm{iso}} + \frac{(13 + \cos^2\theta)}{3} H_{\mathrm{V}} \right\} \sin\theta \; d\theta. \tag{A13}$$

Furthermore, the intensity of the isotropic light scattering, $V_{\mathrm{V}}^{\mathrm{iso}}$, and the anisotropic light scattering, H_{V}, can be expressed by Equation A14 [35] and Equation A15 [34], respectively, as

$$V_{\mathrm{V}}^{\mathrm{iso}} = \frac{\pi^2}{9\lambda_0^4}(n^2 - 1)^2(n^2 + 2)^2 kT\beta, \tag{A14}$$

$$H_{\mathrm{V}} = \frac{16\pi^4}{135\lambda_0^4}(n^2 + 2)^2 N\langle\delta^2\rangle. \tag{A15}$$

Here, λ_0 is the wavelength of light in vacuum, n is the refractive index, k is the Boltzmann constant, T is the absolute temperature, β is the isothermal compressibility, N is the number of scattering units per unit volume, and $\langle\delta^2\rangle$ is the mean square of the anisotropic parameter of polarizability per scattering unit. Finally, from the definition of the turbidity τ in Equation A8, the light scattering loss α_{s} (dB/km) is related to the turbidity τ (cm^{-1}) by

$$\alpha_{\mathrm{s}} = 4.342 \times 10^5 \tau. \tag{A16}$$

Through analyses of light scattering in several optical polymers such as PMMA [36, 37], PSt [38], and PC [39, 40], these equations have proven to be suitable for amorphous polymers with heterogeneous structures small in size relative to the wavelength of the incident light. In particular, PMMA, which has the lowest scattering loss among such polymers, has received considerable attention. Using published data of $\beta = 3.55 \times 10^{-11}\ \mathrm{cm^2/dyn}$ at around T_{g} for bulk PMMA [41] and assuming freezing conditions, the value of $V_{\mathrm{V}}^{\mathrm{iso}}$ at room temperature and a wavelength of 633 nm is $2.61 \times 10^{-6}\ \mathrm{cm}^{-1}$. As a result, the theoretical light scattering loss can be estimated from the $V_{\mathrm{V}}^{\mathrm{iso}}$ value as 9.5 dB/km by using Equations A13 and A16; this is almost identical to the experimental value of 9.7 dB/km. It should be noted here that even in highly purified PMMA, if it is polymerized below the glass transition temperature (T_{g}), the scattering loss inevitably increases to several hundred decibels per kilometer [36]. This is because of the large-sized heterogeneities with dimensions of approximately 1000 Å formed by volume shrinkage during polymerization.

A.4
Graded-Index Technologies for Faster Transmission

After the studies on transparency, a complete high-speed optical network based on POFs became realistic. However, despite these advances, the SI POF had a limitation in bandwidth, requiring the excitation of several tens of thousands of modes for transmission. The large core increases modal dispersion and drastically degrades the bandwidth to approximately several hundred megahertz over 100 m. The concept of fiber bandwidth originates from the general theory of time-invariant linear systems [42]. If the optical fiber can be treated as a linear system, its input and output powers in the time domain are described simply as follows:

$$p_{\text{out}}(t) = h(t) * p_{\text{in}}(t) = \int_{-\frac{T}{2}}^{\frac{T}{2}} p_{\text{in}}(t-\tau)h(\tau)d\tau. \quad \text{(A17)}$$

In other words, the output pulse response $p_{\text{out}}(t)$ of the fiber can be calculated through the convolution of the input pulse $p_{\text{in}}(t)$ and the impulse response function $h(t)$ of the fiber. The period T between input pulses should be wider than the expected time spread of the output pulses. In the frequency domain, Equation A17 can be expressed as follows:

$$P_{\text{out}}(f) = H(f)P_{\text{in}}(f). \quad \text{(A18)}$$

Here, $H(f)$, which is the power transfer function of the fiber at the baseband frequency f, is the Fourier transform of $h(t)$

$$H(f) = \int h(t)e^{-j2\pi ft}dt, \quad \text{(A19)}$$

and $P_{\text{out}}(f)$ and $P_{\text{in}}(f)$ are the Fourier transforms of the output and input pulse responses $p_{\text{out}}(t)$ and $p_{\text{in}}(t)$, respectively, that is,

$$P(f) = \int_{-\infty}^{\infty} p(t)e^{-j2\pi ft}dt. \quad \text{(A20)}$$

The optical bandwidth of the fiber is defined by the Fourier-transformed $H(f)$. This is normally done in terms of the −3-dB bandwidth, which is the modulation frequency at which the optical power of $H(f)$ falls to one-half the value of the zero frequency modulation. The larger the −3-dB bandwidth $f_{-3\,\text{dB}}$, the narrower the output pulse and the higher the possible transmission capacity.

Typically, modal dispersion is the dominant factor degrading the bandwidth for multimode optical fibers such as SI POFs. However, as mentioned above, it can be minimized by forming a quadratic refractive index profile in the core region and controlling the propagation speed of each mode. This method was first developed by Nishizawa *et al.* [13, 16] for glass fibers, and subsequently it was adapted to POFs. The first GI POF was reported in 1982 [43]. Initially, the GI profile was formed by copolymerizing two or three kinds of monomers with different refractive indices and monomer reactivity ratios [44, 45]. In this system, several properties such as attenuation, bandwidth, refractive index profile, and numerical aperture were limited by the differences in the refractive indices and reactivity ratios between the monomers. To solve the problem, a simpler method of forming the GI profile – the low-molecular doping method – was developed in 1994 [46]. By adding a dopant with a higher refractive index than the base polymer and forming a concentration distribution in the radial direction, it became easy to control the refractive index profile. Furthermore, this groundbreaking method enabled the reliable fabrication of low-loss GI POFs. Since the first GI POF was reported, various methods for forming a GI profile have been proposed [47–52]. Among these, here we discuss two major techniques using the low-molecular doping method.

A.4.1
Interfacial-Gel Polymerization Technique

In this method, a rod with a GI profile called a *preform* is first prepared, after which the preform is heat-drawn to the GI POF. The interfacial-gel polymerization technique [53] refers to the method used to form a parabolic refractive index profile in the core region. In this review, we use a typical system as an example. The base polymer of the core and cladding layers is PMMA and the low-molecular-weight dopant is diphenyl sulfide (DPS). First, a glass tube charged with MMA monomer mixtures including benzoyl peroxide (BPO) and *n*-butyl mercaptan (*n*-BM) as the initiator and chain-transfer agent, respectively, is rotated on its axis at 3000 rpm in an oven at 70 °C for 3–6 h. The MMA monomer gets coated on the inner wall of the glass tube because of centrifugal force, and is gradually polymerized (Figure A.6a). After heat treatment at 90 °C for 24 h, the polymer tube based on PMMA is obtained as the cladding layer of the GI preform. The tube is then filled with a core solution containing a mixture of MMA monomers as well as DPS, di-*tert*-butyl peroxide (DBPO), and *n*-lauryl mercaptan (*n*-LM) as the dopant, initiator, and chain-transfer agent, respectively. The tube filled with the mixture is heated in an oil bath at 120 °C for 48 h under a nitrogen pressure of 0.6 MPa (Figure A.6b). Previous studies have shown that an amorphous polymer glass such as PMMA polymerized at a temperature lower than the T_g exhibits an extremely high scattering loss because of density fluctuations [36–38]. Hence the core solution has to be polymerized at a temperature higher than the T_g, which is 105–115 °C. However, such a high temperature leads to numerous bubbles during the polymerization reaction because it is above the boiling point of MMA (100 °C). Therefore, polymerization must be performed under an appropriate pressure. Finally, the preform is heat-drawn to the fiber at 220–250 °C (Figure A.6c).

The mechanism for forming a GI profile is described in Figure A.7. At the beginning of core polymerization, the inner wall of the PMMA tube becomes slightly swollen by the monomer–dopant mixture to form a polymer gel phase. The reaction rate of polymerization is generally faster in the gel phase than in the monomer

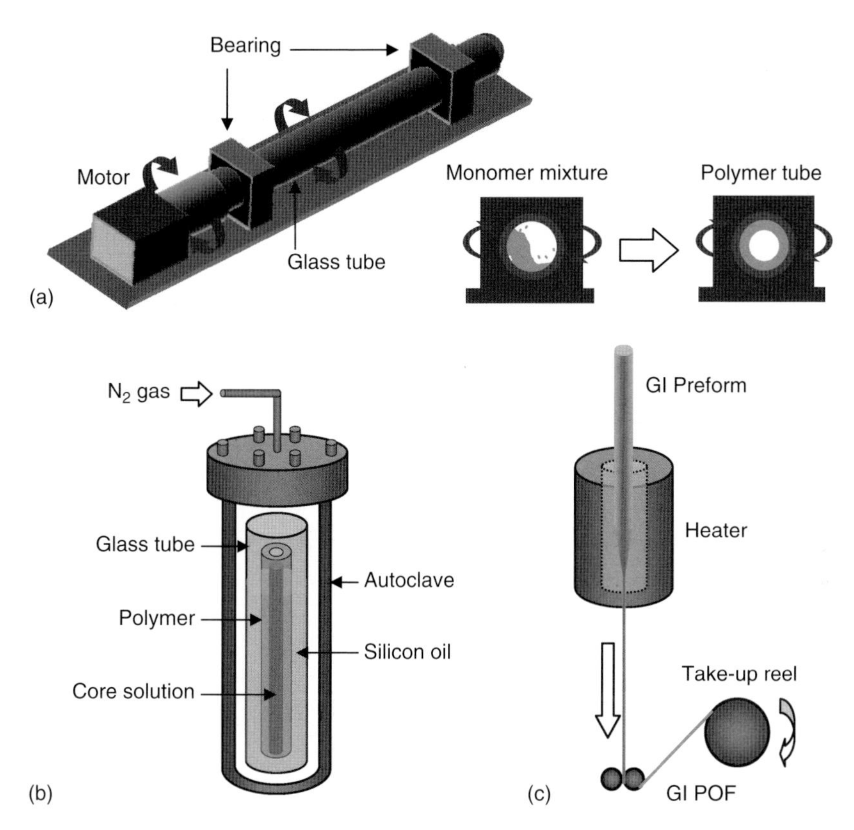

Figure A.6 Fabrication process of GI POFs by the preform method. (a) Preparation of polymer tube. (b) Core polymerization. (c) Heat-drawing.

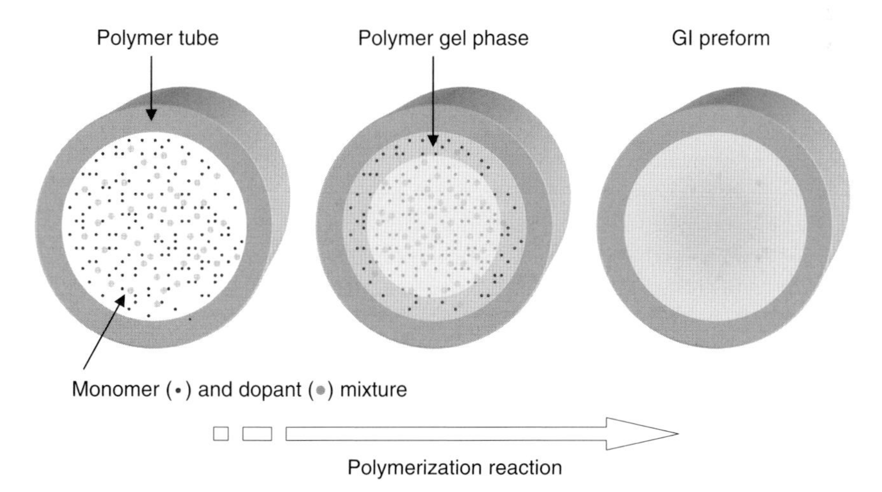

Figure A.7 Formation process of GI distribution by the interfacial-gel polymerization technique.

mixture because of the gel effect [54], and the polymer phase grows from the inner wall of the tube to the core center. During the growth process, the MMA monomer can diffuse into the gel phase more easily than the dopant molecules because the molecular volume of the dopant, which contains benzene rings, is larger than that of the monomer. Hence, the dopant is concentrated in the middle region to form a quadratic refractive index profile. The shape of the profile can be adjusted by various methods, and precise control is possible [55]. Above all, the amount of initiator, the polymerization temperature, and the ratio of the diameter of the hole to that of the tube strongly affect the profile.

The interfacial-gel polymerization technique is particularly common in acrylic GI POF studies and enables the precise control of the refractive index profile, leading to a maximal bandwidth. However, this batch process requires many complicated procedures. Furthermore, the fiber length obtained at any one time is completely dependent on the preform size. This is a serious limitation in terms of fabrication costs.

A.4.2
Coextrusion Process

In order to realize mass production, a new process for GI POF fabrication called *coextrusion* was investigated [56–58]. In this process, polymers doped with low-molecular-weight dopants and homogeneous polymers are first prepared as the base materials of the core and cladding layers, respectively. In PMMA-DPS systems, polymerizations are performed at 90 °C for 24 h. The bulk polymers are then further heated at 110 °C for 48 h. The two materials are melted in their respective extrusion sections at 210–250 °C and compounded in a die to fabricate a POF with a concentric circular core/cladding structure. At this point, the fiber does not have a GI distribution. However, by heating the fiber in the diffusion section, a radial concentration profile of the low-molecular-weight dopant is formed as a result of molecular diffusion. Finally, the GI POF is obtained by winding it onto a take-up reel. The operational procedure of coextrusion is illustrated in Figure A.8. This process is very simple, and hence GI POFs are expected to be manufactured continuously at a relatively low cost.

In previous studies, it was presumed that a GI POF fabricated by this process would have a refractive index distribution with a tail at the core/cladding boundary, not as steep as that prepared by the conventional batch process [50]. Therefore, the bandwidth of the GI POF obtained by dopant diffusion coextrusion would be inferior to that obtained by the batch process. However, contrary to this expectation, a GI POF without any tail prepared by the process was reported in 2007 [57]. If low-molecular-weight molecules diffuse in a radial direction, their diffusion can be expressed by Fick's diffusion equation

$$\frac{\partial C}{\partial t} = \frac{1}{r}\frac{\partial}{\partial r}\left(Dr\frac{\partial}{\partial r}C\right). \tag{A21}$$

Here, C is the concentration of low-weight molecules, t is the diffusion time, r is the distance from the fiber center, and D is the mutual diffusion coefficient of

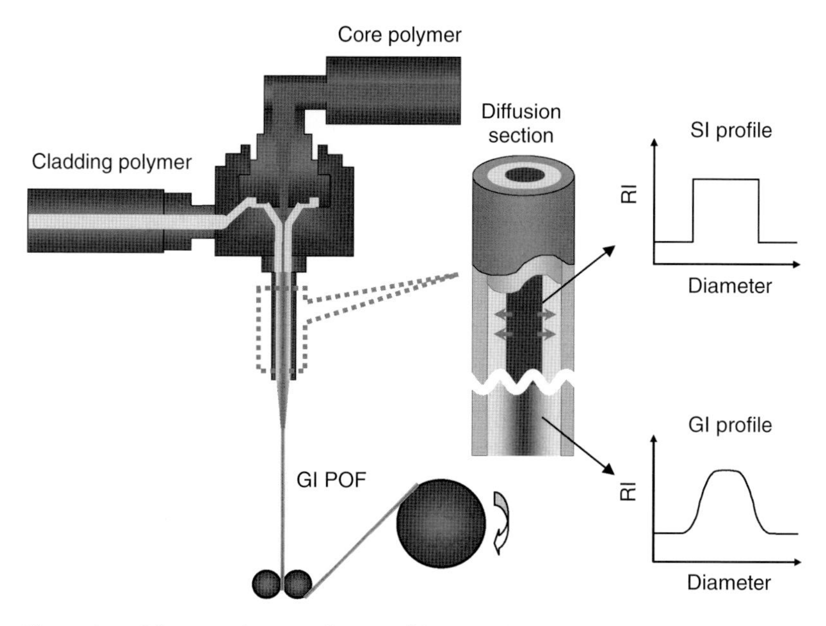

Figure A.8 Schematic diagram of GI POF fabrication by the coextrusion procedure.

low-weight molecules. The concentration profile calculated using Equation A21 has a gradually varying shape around the diffusion front because D is almost constant in general systems. However, in the polymer–dopant system, the plasticization that occurs as a result of adding the dopant into the polymer matrix must be considered. As the dopant concentration increases, T_g, which determines the mobility and diffusivity of the dopant, decreases linearly. In other words, diffusion within the diffusion section of the coextrusion equipment obeys Fick's law with a diffusion coefficient that varies as a function of the dopant concentration. Figure A.9 shows the comparison of the refractive index profiles of experimental and simulated data. The simulation was performed with a diffusion coefficient calculated from the results of a one-dimensional diffusion experiment. The polymer matrix and dopant are PMMA and DPS, respectively. The experimental and simulated data show good agreement, indicating a high possibility that a high-bandwidth GI POF can be prepared by the dopant diffusion coextrusion process. Currently, this continuous fabrication method is well established and is employed for some commercially available GI POFs.

A.5
Recent Studies of Low-Loss and Low-Dispersion Polymer Materials

With various improvements in the fabrication process enabling the avoidance of contaminants and the formation of GI profiles, low-loss and high-bandwidth POFs have become leading candidates for short-distance networks (e.g.,

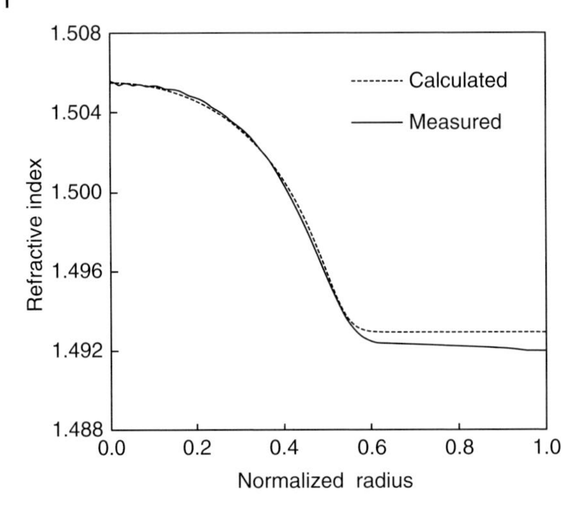

Figure A.9 Refractive index profiles. Broken line is the refractive index profile calculated from Equation A21 using the measured dependence of the mutual diffusion coefficient for a PMMA-DPS system. Solid line is the measured refractive index profile for a PMMA-DPS-based GI POF fabricated by the coextrusion method. (Adapted with the permission from Ref. [57], © 2007 IEEE.)

home networks). For practical applications, however, the intrinsic problem of attenuation still remained.

Since the first SI POF was commercialized, most POFs have been manufactured using PMMA, a mass-produced, commercially available polymer that demonstrates high light transmittance and provides excellent corrosion resistance to both chemicals and weather. These properties, coupled with low manufacturing costs and easy processing, have made PMMA a valuable substitute for glass in optical fibers. PMMA-based SI POFs have been used extensively in short-distance datacom applications such as digital-audio interfaces. They have also been utilized for data transmission equipment, control signal transmissions for numerically controlled machine tools, railway rolling stocks, and optical data buses in automobiles. In particular, the increasing complexity of in-vehicle electronic systems has led to POFs becoming indispensable to the automobile industry [59]. Today, it is not uncommon to find 10 or 20 consumer electronic devices, such as main units, DVD (blue-ray) players, navigation systems, telephones, Bluetooth interfaces, voice recognition systems, high-end amplifiers, and TV tuners, all connected inside a car. To meet all the necessary requirements for data transfer between such devices, POFs have provided a great solution.

On the other hand, when high-speed data transmission of more than 1 Gbps is required, PMMA-based POFs cannot be used. In the short-distance datacom or telecom applications above, systems using SI POFs as the transmission medium and red light-emitting diodes (LEDs) at 650 nm as the light source have been employed. In contrast, to realize gigabit in-home communications, GI POFs along with vertical-cavity surface-emitting lasers (VCSELs) [60], which offer fast modulation, form the most reasonable combination [61]. VCSELs constitute a relatively new class of semiconductor lasers which are monolithically fabricated. They are now considered to be key devices for Gigabit Ethernet, high-speed LANs (local area networks), computer links, and optical interconnects. As mentioned before, an acceptable level of fiber attenuation in home networks is ∼200 dB/km. Furthermore, since the emission wavelengths of long-life and inexpensive VCSELs are 670–680 nm, the fiber must satisfy the limitations at this wavelength region. However, for PMMA-based POFs, the wavelengths satisfying the requirement are limited to 570 and 650 nm because of absorption losses from C–H stretching vibrations [38].

Another widely used base polymer for POF is PSt. With respect to mechanical properties and chemical resistance, PSt is slightly inferior to PMMA. However, PSt has some features appropriate for the core material of POF, as well as an attractive price. Among them, the biggest advantage of using PSt is the low attenuation at wavelengths of 670–680 nm. While PMMA-based POFs have a high attenuation of over 200 dB/km in this region, the attenuation of PSt-based POFs is as low as 114 dB/km [27]. Although the visible spectrum of a PSt-based POF is dominated by high harmonic C–H absorptions as in the case of a PMMA-based POF, the shape of the spectrum is completely different. The attenuation spectrum of a PSt-based POF is shown in Figure A.10. In the styrene unit, there are three aliphatic and five aromatic C–H bonds. The absorption wavelengths of the aliphatic C–H bonds corresponding to the fifth, sixth, and seventh overtones are 758, 646, and

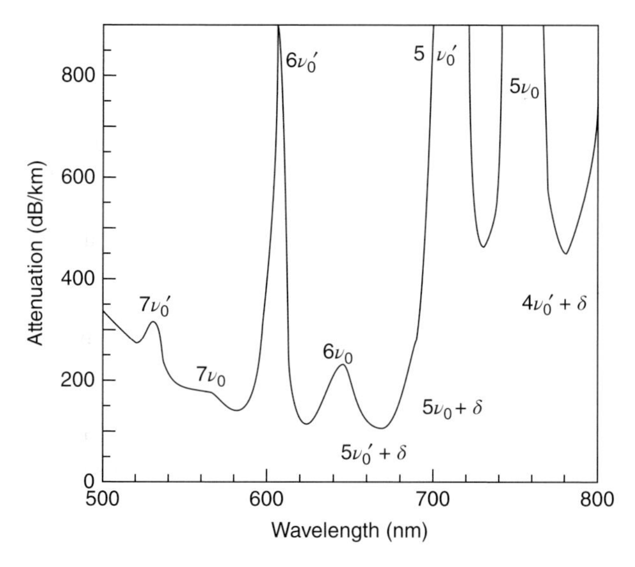

Figure A.10 Attenuation spectrum of PSt-based SI POF. ν_0 and ν_0' correspond to absorption wavelengths of aliphatic and aromatic C–H bonds, respectively. Shoulders in the region of lower wavelength of each peak are assigned to the combination bands $n\nu_0 + \delta$ and $n\nu_0' + \delta$, where $+\delta$ is the fundamental aliphatic C–H bending vibration. (Adapted with permission from Ref. [27], © 1981 AIP.)

562 nm, respectively, whereas the same overtones of the aromatic C–H bonds appear at 714, 608, and 532 nm, respectively [27, 32]. It is noteworthy that the positions of the aliphatic C–H overtones are slightly shifted to longer wavelengths relative to those for PMMA. As a result, the emission wavelength of a VCSEL is centered between the fifth aromatic and sixth aliphatic C–H peaks and is not significantly affected by C–H absorption losses. Most studies on PSt-based POFs have considered SI-type POFs [30, 32]. If a high-speed GI POF based on PSt can be fabricated with low attenuation, the fiber could be a highly promising candidate for a home network medium. However, this is not easy. The difficulty arises from the high refractive index. The refractive index of PSt is around 1.59, which is higher than that of PMMA by 0.1. The importance of this difference can be observed in Figure A.9. When a GI profile with a sufficient difference in refractive indices between the core and cladding is formed by doping DPS, the T_g of the core center decreases to below 60 °C because of the plasticization effect. Clearly, the thermal stability is insufficient for practical use.

A.5.1
Partially Fluorinated Polymers

The high attenuation of conventional POFs is dominated by C–H overtone stretches and combinations of stretch and deformations. Hence, the most effective method to obtain a low-loss POF is substituting hydrogen with heavier atoms such as fluorine, as was shown in Figure A.5. However, if vinyl monomers such as MMA are perfluorinated, the polymerization rate drastically decreases. Boutevin *et al.* studied various partially fluorinated polymers which are easily prepared by free-radical polymerization and calculated the influence of the molar number of C–H bonds per unit volume [62]. The contribution of the fluorine substituent is considerably larger than expected from the chemical structure. For instance, poly(2,2,2-trifluoroethyl methacrylate) (poly(TFEMA))-based GI POF provides a surprisingly low attenuation of 127–152 dB/km at 670–680 nm [63] relative to that for PMMA-based GI POF. Although the number of C–H bonds per monomer unit between TFEMA and MMA differs by only 1, the trifluoroethyl group of TFEMA possesses a large volume, and the number of C–H bonds per unit volume in poly(TFEMA) is only 64% of that in PMMA. However, the long side chain not only reduces fiber attenuation but also lowers T_g.

Recently, a novel GI POF based on a partially fluorinated polymer with a lower attenuation and higher T_g than those of PMMA-based GI POF was reported [64]. In this study, a copolymer of MMA and pentafluorophenyl methacrylate (PFPMA) was employed as the core polymer. In general, copolymers tend to have extremely high scattering losses because of their large heterogeneous structure and corresponding heterogeneity of the refractive index; consequently, they have not been leading candidates for POF base materials. In this copolymeric system, however, the increase in scattering loss is negligibly small since the refractive indices of both homopolymers are almost identical (PMMA: $n_D = 1.4914$, poly(PFPMA): $n_D = 1.4873$). Moreover, the C–H bonds per unit

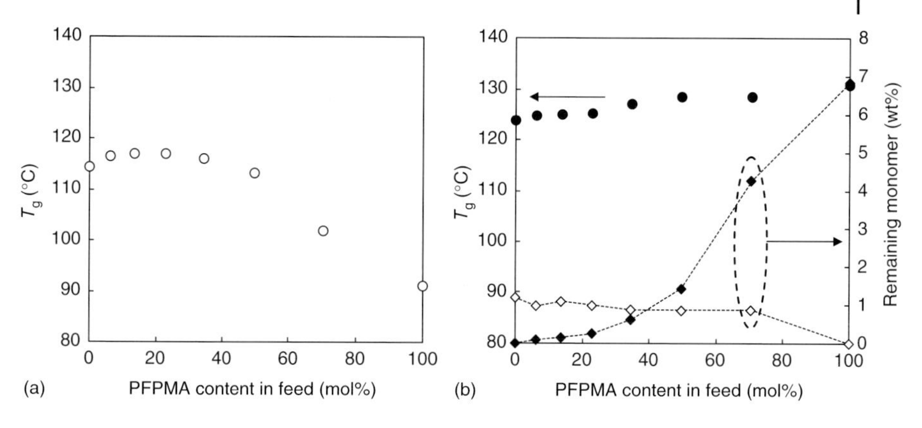

Figure A.11 (a) T_g plots of bulk MMA-*co*-PFPMA polymerized at 110 °C for 48 h (○). (b) T_g plots of the purified MMA-*co*-PFPMA (•), and the amount of residual MMA (◊) and PFPMA (♦) in bulk copolymers against the PFPMA content in the monomer feed. (Adapted with permission from Ref. [64], © 2010 Elsevier Ltd.)

volume in poly(PFPMA) compared to PMMA is only 34%; hence, smaller absorption losses and lower attenuation than those for PMMA can be obtained by copolymerization. When the core composition is 65 : 35 of MMA/PFPMA (mol%), the attenuations are 172–185 dB/km at 670–680 nm; this satisfies the required attenuation for optical home network systems.

Regarding thermal stability, the copolymer exhibits a higher T_g than PMMA when the PFPMA content in the monomer feed is 0–50 mol%. The relationship between the T_g of bulk MMA-*co*-PFPMA and the PFPMA content in the monomer feed is shown in Figure A.11a. The T_g for copolymers can generally be described by the Gordon–Taylor [65] equation and changes linearly with the weight fractions of the monomer content. Interestingly, in the case of this copolymeric system, it is observed that the T_g plot exhibits a positive deviation. Although the abscissa is expressed as the molar fraction of PFPMA in the monomer feed, the result is similar even if it is converted to weight fractions. This result can be explained by examining Figure A.11b, in which the T_g of precipitated copolymers and the amount of residual monomers in the bulk copolymer are shown. As Boutevin *et al.* reported, poly(PFPMA) has an intrinsically high T_g of 130 °C, whereas the bulk shows a low T_g of 91 °C. The reactivity of the carbon–carbon double bond of phenyl methacrylate is known to be significantly reduced by substituting the hydrogen of the benzene ring with larger molecules [66, 67]. In this case, the pentafluorophenyl group disturbs the propagation reaction with steric hindrance, thereby leading to low polymerization conversion. As a result, numerous unreacted monomers remain in the bulk, and the T_g is degraded by the plasticization effect. That the T_g of the copolymer increases as the PFPMA content decreases is related to changes in the amount of PFPMA residues.

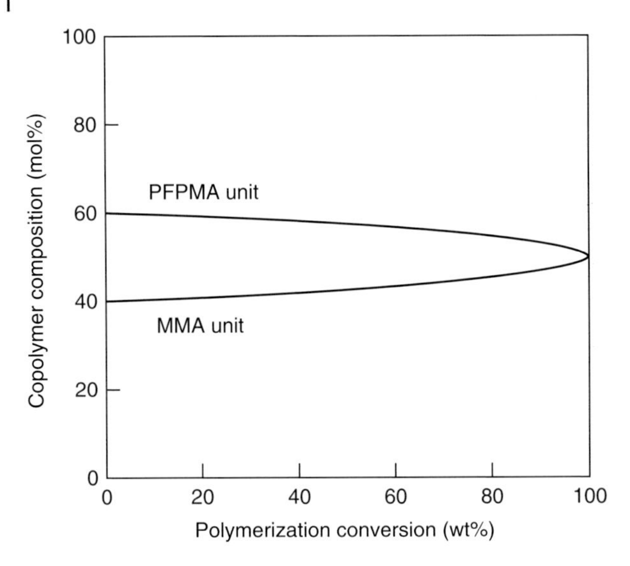

Figure A.12 Change in accumulated copolymer composition with polymerization conversion when the initial MMA and PFPMA contents in the monomer feed are 50 mol%. (Adapted with permission from Ref. [64], © 2010 Elsevier Ltd.)

The amount of MMA remaining in all bulks except poly(PFPMA) is ~1 wt% and nearly independent of the monomer feed composition. In contrast, the residual PFPMA drastically decreases after copolymerization with MMA. This derives from the monomer reactivity ratios. Figure A.12 shows the changes in the copolymer composition with the polymerization conversion calculated by the Mayo–Lewis equation. Here, r_{12} and r_{21} (M_1: MMA, M_2: PFPMA) are 0.56 and 1.30, respectively [68], and the initial monomer content in the feed is 50/50 mol%. As shown in the figure, PFPMA polymerizes preferentially, assuming ideal polymerization. As the polymer initially consists of 60 mol% PFPMA units, PFPMA conversion can be improved in the presence of MMA. On the other hand, the T_g of the purified copolymer with no residual monomers linearly increases with the PFPMA content in the monomer feed because of the high T_g of poly(PFPMA). Therefore, the T_g of the bulk copolymer increases with increasing PFPMA content when the bulk has a small amount of residual monomers, whereas it decreases when the bulk has numerous residual monomers because of the plasticization effect. This is why the T_g plots for the bulk copolymers exhibit a positive deviation. The most important aspect of these experiments is that the T_g of the bulk copolymers is sufficiently high, and that the bulks – across a wide range of copolymer compositions – can be used as the base material for GI POFs as a transmission medium in home networks. In the copolymer study, both the core and cladding layers were designed to have good stability against humidity, and the minimum T_g of a GI POF was controlled to be over 90 °C, ensuring long-term usage in home environments.

A.5.2
Perfluorinated Polymer

For further reduction of fiber attenuation, perfluorinated polymers have also been intensively investigated. Compared to a large number of radically polymerizable hydrocarbon monomers, only a few classes of perfluoromonomers can homopolymerize under normal conditions via the free-radical mechanism. The most typical example of a perfluoromonomer that can be prepared via free-radical polymerization is tetrafluoroethylene (TFE), developed by DuPont in 1938. As is well known, poly(TFE) (Teflon™) is opaque despite the lack of C–H bonds. In general, perfluorinated resins are rigid and easily form partially crystalline structures [69]. Hence, light is scattered at the boundary between the amorphous and crystalline phases, causing haziness. To avoid the crystalline form, an effective method is to introduce aliphatic rings into the main chain, which becomes twisted and incapable of forming a crystalline structure. The most famous examples of amorphous perfluorinated polymers are Teflon AF and CYTOP™ [70], which were developed by DuPont and AGC, respectively. Their chemical structures are shown in Figure A.13. Both have excellent clarity, solubility in fluorinated solvents, thermal and chemical durability, high electrical isolation, low water absorption, and low dielectric properties. In particular, their high transparencies arise from the cyclic structures existing in the polymer main chains. Teflon AF is a copolymer of perfluoro-2,2-dimethyl-1,3-dioxole (PDD), which possesses a cyclic structure in its monomer unit, and TFE. On the other hand, CYTOP is a homopolymer of perfluoro(4-vinyloxyl-1-butene) (BVE), and cyclopolymerization yields the cyclic structures (penta- and hexa-cyclic) on the polymer chain [71].

Although both have sufficient properties as the core material, Teflon AF has mainly been utilized as the cladding layer of optical fibers or waveguides because

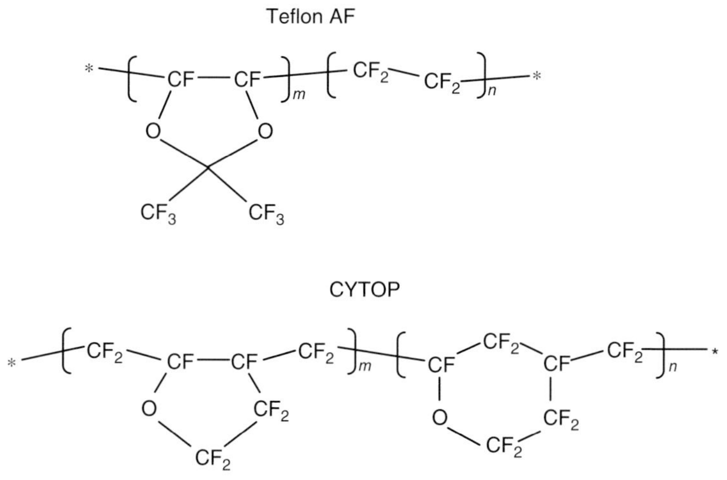

Teflon AF

CYTOP

Figure A.13 Chemical structures of Teflon AF and CYTOP.

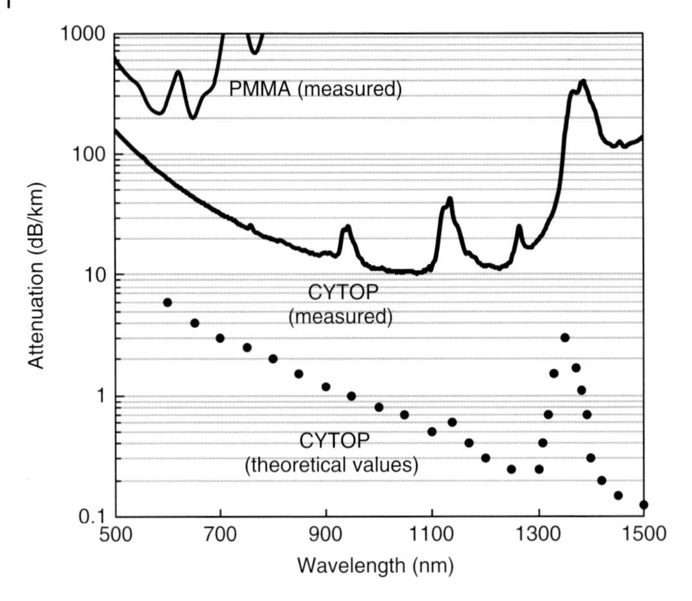

Figure A.14 Comparison of attenuation spectra between PMMA- and CYTOP-based GI POFs. (Adapted with permission from Ref. [55], © 2006 IEEE.)

of its extremely low refractive index ($n_D = 1.29$). Given that this value is lower than even that of water, liquid-core optical fibers (LCOFs) have been studied for other purposes such as Raman spectroscopy analysis [72, 73]. In 2000, AGC commercialized the first CYTOP-based GI POF (Lucina™), which has been adopted by condominiums, hospitals, data centers, and other facilities in Japan [74]. Figure A.14 is a comparison of the attenuation spectra for PMMA- and CYTOP-based GI POFs. CYTOP molecules consist solely of C–C, C–F, and C–O bonds. The wavelengths of the fundamental stretching vibrations of these atomic bonds are relatively long; therefore, vibrational absorption loss of CYTOP at the region of the light source wavelength is negligibly small. In addition, it shows fairly low light scattering because of the low refractive index ($n_D = 1.34$) (see Equation A14). The attenuation of CYTOP-based GI POF is ~10 dB/km at 1.0 μm. Given that the theoretical limit of attenuation is 0.7 dB/km at this wavelength [75], it is expected that the attenuation can be lowered further by preventing contamination during the fabrication process. The reason why the attenuation at 670–680 nm is slightly higher than that at around 1.0 μm is that the intrinsic scattering loss expressed by Equation A14 is inversely proportional to λ^4. In this case, the effect of the anisotropic scattering expressed by Equation A15 need not be considered because the anisotropy of the polarizability in CYTOP is negligibly small.

In fact, the excellent low-loss characteristic of CYTOP-based GI POF (e.g., 10 dB/km) is far beyond the requirement of home networking. However, its real uniqueness is the low material dispersion derived from the low refractive index. In the past, there have been several reports on perdeuterated PMMA (PMMA-d_8)

as a novel base material for low-loss POFs [28, 76, 77]. The replacement of deuterium for hydrogen in PMMA also results in considerable reduction in the C–H vibrational absorptions in the infrared region and in its overtones in the visible-to-near infrared region. As a result, loss reductions of 20 and 63 dB/km at 670–680 nm for SI POF [76] and GI POF [77], respectively, have been successfully achieved; however, the substitution of hydrogen with deuterium does not lower the refractive index and material dispersion. The bandwidth of optical fibers that excite many modes (e.g., POFs) is predominantly influenced by modal dispersion. However, once the modal dispersion is reduced by forming a GI profile, the influence of material dispersion on bandwidth can no longer be ignored. Material dispersion is induced by the wavelength dependence of the refractive index of the fiber material and the finite spectral width of the light source.

The refractive index profiles of GI POFs can be approximated by the following power law [78]:

$$\begin{aligned} n(r) &= n_1\left[1 - 2\Delta\left(\frac{r}{R}\right)^g\right]^{\frac{1}{2}}, \quad && 0 \leq r \leq R \\ &= n_1(1 - 2\Delta)^{\frac{1}{2}}, && R < r. \end{aligned} \tag{A22}$$

Here, $n(r)$ is the refractive index n at radius r of the fiber, n_1 is the refractive index at the core center, n_2 is the refractive index of the cladding, $R$is the core diameter, $g$is the refractive index profile coefficient, and $\Delta$is the relative index difference defined as

$$\Delta = \frac{n_1{}^2 - n_2{}^2}{2n_1{}^2} \cong \frac{n_1 - n_2}{n_1}. \tag{A23}$$

The crucial factor defining the refractive index profile in GI POFs is the coefficient g, and the optimum value for maximizing the bandwidth can be determined from the modal and material dispersions [79–81]. From analyses using the Wentzel–Kramers–Brillouin (WKB) method, the modal dispersion σ_{inter}, material dispersion σ_{intra}, and total dispersion σ_{total} can be expressed as follows:

$$\sigma_{\text{inter}} = \frac{Ln\Delta}{2c}\frac{g}{g+1}\left(\frac{g+2}{3g+2}\right)^{\frac{1}{2}}\left[S_1^2 + \frac{4S_1S_2\Delta\,(g+1)}{2g+1} + \frac{4S_2^2\Delta^2(2g+2)^2}{(5g+2)(3g+2)}\right]^{\frac{1}{2}}, \tag{A24}$$

$$S_1 = \frac{g-2-\varepsilon}{g+2}, \quad S_2 = \frac{3g-2-2\varepsilon}{2(g+2)}, \quad \varepsilon = \frac{-2n_1}{n}\frac{\lambda d\Delta}{\Delta d\lambda}, \quad n = n_1 - \lambda\frac{dn_1}{d\lambda}, \tag{A25}$$

$$\begin{aligned} \sigma_{\text{intra}} &= \frac{L\sigma_s}{\lambda}\left[\left(-\lambda^2\frac{d^2n_1}{d\lambda^2}\right)^2 - 2\lambda^2\frac{d^2n_1}{d\lambda^2}(n\Delta)S_1\left(\frac{2g}{2g+2}\right)\right. \\ &\quad \left. + (n\Delta)^2\left(\frac{g-2-\varepsilon}{g+2}\right)^2\frac{2g}{3g+2}\right]^{\frac{1}{2}}, \end{aligned} \tag{A26}$$

$$\sigma_{\text{total}} = \sqrt{\sigma_{\text{inter}}^2 + \sigma_{\text{intra}}^2}. \tag{A27}$$

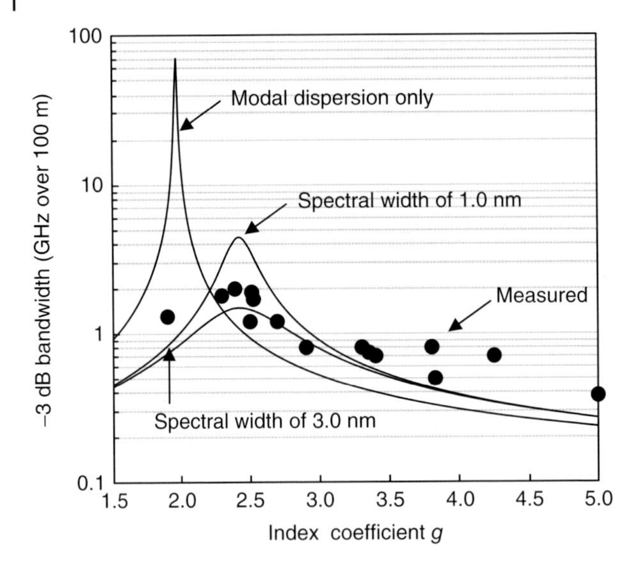

Figure A.15 Relationship between index profile coefficient g and the −3-dB bandwidth over 100 m of PMMA-DPS-based GI POF at a wavelength of 650 nm. Solid lines are the calculated results. Closed circles are the measured bandwidths of GI POFs prepared by the interfacial-gel polymerization technique (spectral width is 3.0 nm). (Adapted with permission from Ref. [55], © 2006 IEEE.)

Here, the spectral width σ_s is the wavelength dispersion of the input pulse, $c$is the velocity of light, $\lambda$is the wavelength of light, and $L$is the transmission distance. Before turning to the material dispersion of CYTOP, we first consider the relationships among the refractive index coefficient g, modal dispersion, material dispersion, and the −3-dB bandwidth by using a typical example. Figure A.15 shows the relationship between the theoretical bandwidth and the index exponent g for GI POFs based on PMMA at an operating wavelength of 650 nm. The theoretical bandwidth was calculated using the wavelength dispersion of the light source at $\sigma_s = 1.0$ and 3.0 nm with a fiber length equal to 100 m. Figure A.15 indicates that the theoretical limit of the bandwidth is ~3.0 GHz over 100 m when the spectral width of the light source is 1.0 nm. When the index exponent g differs from the optimum value, the spectral width dependence of the bandwidth is negligibly small since σ_{inter} is considerably larger than σ_{intra}. However, when g ranges between 2 and 3, there is a significant difference between these two curves; this is the effect of the material dispersion of PMMA and the dopant included in the core. The experimentally measured bandwidth shown by the closed circles is well predicted by taking into account the material dispersion.

Figure A.16 shows a comparison of the material dispersions among PMMA, CYTOP, and silica glass. The material dispersion of CYTOP is even lower than that of silica glass. The theoretical dependence of the bandwidth of the CYTOP-based GI POF and a multimode GOF on wavelength is shown in Figure A.17. CYTOP-based GI POFs are predicted to have a higher bandwidth than multimode GOFs

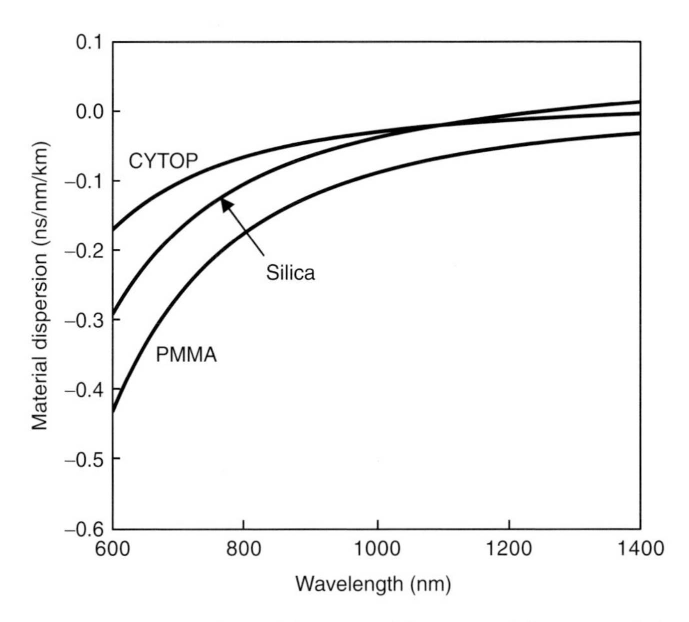

Figure A.16 Comparison of the material dispersion of fiber materials for pure CYTOP, PMMA, and silica. (Adapted with permission from Ref. [55], © 2006 IEEE.)

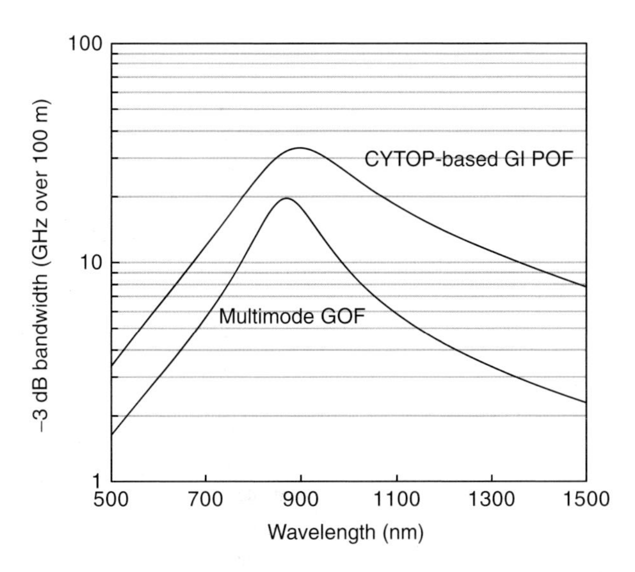

Figure A.17 Dependence of the theoretical −3-dB bandwidth on wavelength in the CYTOP-based GI POF link compared to that in a multimode GOF ($\sigma_s = 1.0$ nm). Bandwidth is calculated on the assumption that each fiber has an optimal refractive index profile at 850 nm. (Adapted with permission from Ref. [55], © 2006 IEEE.)

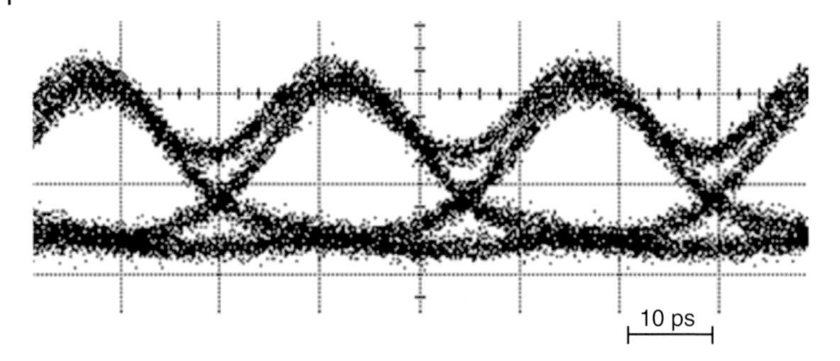

Figure A.18 Eye diagram of 40 Gbps data transmission at 1550 nm propagated through the 100-m GI POF based on CYTOP.

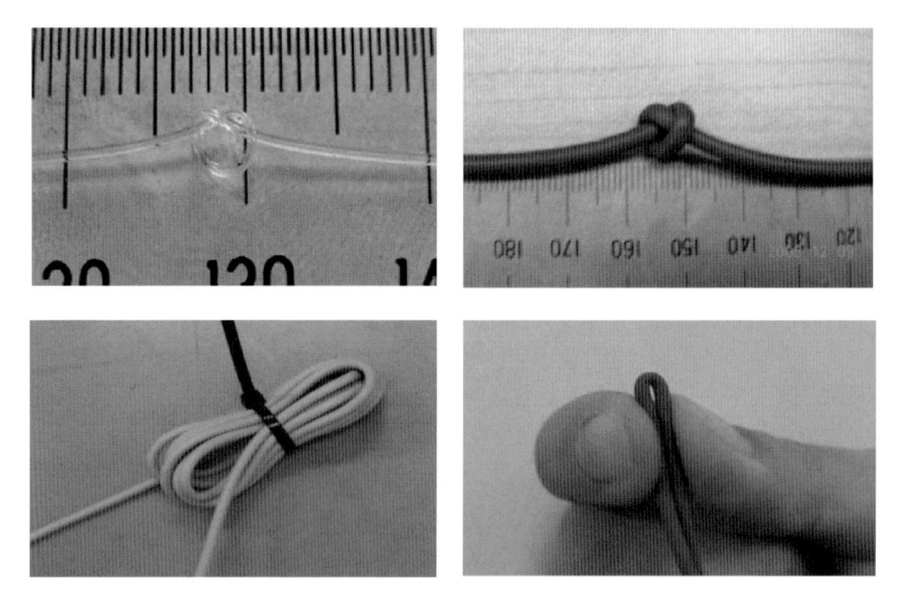

Figure A.19 Photographs of the double-cladding CYTOP-based GI POF (Fontex) provided by AGC.

[55]. Indeed, in 1999, AGC and Bell Laboratories reported successful experimental transmission at 11 Gbps over 100 m using CYTOP-based GI POFs [82]. Furthermore, CYTOP-based GI POFs capable of 40 Gbps transmission over 100 m were subsequently reported in 2007 [83] and 2008 [84, 85]. Figure A.18 is the eye pattern of 40 Gbps data transmission through a CYTOP-based GI POF. The eye-pattern method is described in greater detail elsewhere. Suffice it to say that the good eye opening clearly shows that this fiber sufficiently ensures 40 Gbps transmission. These are highly significant results and demonstrate that POFs can achieve a higher bandwidth than multimode GOFs.

Recently, demands for the replacement of electrical with optical signals are rapidly increasing not only for in-home networks but also for those of shorter reach, such as in-device and on-board networks. In these areas, the required bit rate is forecast to be tens of gigabits per second in the next decade [86, 87]. Although multimode GI GOFs are currently considered for the transmission medium, the above data show that POFs can offer better performance even in such extremely high speed connections. Moreover, as we have mentioned several times before, POFs can also provide ease of handling and flexibility in the wiring design with its features of plastic-specific robustness. In 2010, AGC released another CYTOP-based GI POF called Fontex™ [88]. By employing a double cladding structure – a thin layer with considerably lower refractive index placed around the first cladding – the bending loss was further reduced while high-speed capacity was maintained [89], thereby enabling various wiring designs to become possible (Figure A.19). In addition, the continuous fabrication of GI POF has also been established by the coextrusion process.

A.6 Conclusion

This appendix reviewed the status of POF developments in the last half century, which have focused on loss reduction and bandwidth enhancement. Today, optical fibers are ubiquitous and can be encountered anywhere – in networks such as submarine links, long-distance terrestrial networks, metropolitan and access loops, and FTTH drop architectures. However, they are still not used in home networks. Currently, several technologies are available for broadband home networking. In particular, coaxial cables, twisted-pair cables, and wireless LAN links are being intensively investigated. However, from the standpoints of transmission speed, reliability, ease of handling, and safety, GI POFs appear to be the best solution. Furthermore, POFs are now ready not only for use in home networking but also for device and storage interconnections where data transmissions over 40 Gbps will be required in the near future. POFs are no longer just alternatives to GOFs; their advantages for short-range networks are obvious.

In the future beyond FTTH services, there awaits a world in which all information will be directly transferred by light signals, bringing us back to real-time face-to-face communications with large screens and clear motion pictures. We believe that POF technologies will contribute in no small way to accelerate this paradigm shift.

Acknowledgment

This work was partly supported by the Japan Science and Technology Agency through a grant from the ERATO-SORST Koike Photonics Polymer Project.

References

1. Koike, Y. and Koike, K. (2011) Progress in low-loss and high-bandwidth plastic optical fibers. *J. Polym. Sci., Part B: Polym. Phys.*, **49** (1), 2–17.
2. Wallace, F.J. (1963) *J. Urol.*, **90**, 324–334.
3. Mims, M.M. and Schlumberger, F.C. (1964) *J. Urol.*, **91**, 435–437.
4. Kao, C.K. and Hockham, A.G. (1966) *Proc. IEE*, **113**, 1151–1158.
5. Kapron, F.P., Keck, D.B., and Maurer, R.D. (1970) *Appl. Phys. Lett.*, **17**, 423–425.
6. Izawa, T., Shibata, N., and Takeda, A. (1977) *Appl. Phys. Lett.*, **31**, 33–35.
7. Miya, T., Terunuma, Y., Hosaka, T., and Miyashita, T. (1979) *Electron. Lett.*, **15**, 106–108.
8. Nagayama, K., Matsui, M., Kakui, M., Saitoh, T., Kawasaki, K., Takamizawa, H., Ooga, Y., Tsuchiya, I., and Chigusa, Y. (2004) *SEI Tech. Rev.*, **57**, 3–6.
9. Chanclou, P., Belfqih, Z., Charbonnier, B., Duong, T., Frank, F., Genay, N., Huchard, M., Guignard, P., Guillo, L., Landousies, B., Pizzinat, A., Ramanitra, H., Saliou, F., Durel, S., Urvoas, P., Ouzzif, M., and Masson, J.L. (2008) *C.R. Phys.*, **9**, 935–946.
10. Emslie, C. (1988) *J. Mater. Sci.*, **23**, 2281–2293.
11. White, K.I. and Nelson, B.P. (1979) *Electron. Lett.*, **15**, 394–395.
12. Tsuchiya, H. and Imoto, N. (1979) *Electron. Lett.*, **15**, 476–478.
13. Nishizawa, J. and Sasaki, I. (1964) JP Patent 46 29291.
14. Miller, S.E. (1965) *Bell Syst. Tech. J.*, **44**, 2017–2063.
15. Uchida, T., Furukawa, M., Kitano, I., Koizumi, K., and Matsumura, H. (1970) *IEEE J. Quantum Electron.*, **QE-6**, 606–612.
16. Nishizawa, J. and Otsuka, A. (1972) *Appl. Phys. Lett.*, **21**, 48–50.
17. Eijkelenborg, M.A., Large, M., Argyros, A., Zagari, J., Manos, S., Issa, N.A., Bassett, I., Fleming, S., McPhedran, R.C., Sterke, C.M., and Nicorovici, N. (2001) *Opt. Express*, **9**, 319–327.
18. Kalli, K., Dobb, H.L., Webb, D.J., Carroll, K., Komodromos, M., Themistos, C., Peng, G.D., Fang, Q., and Boyd, I.W. (2007) *Opt. Express*, **32**, 214–216.
19. RongJin, Y. and Bing, Z. (2008) *Sci. China Ser. E: Technol. Sci.*, **51**, 2207–2217.
20. Kong, D. and Wang, L. (2009) *Opt. Lett.*, **34**, 2435–2437.
21. Zhou, G., Pum, C.J., Tam, H., Wong, A.C.L., Lu, C., and Wai, P.K.A. (2010) *IEEE Photonics Technol. Lett.*, **22**, 106–108.
22. DuPont de Nemours & Co (1978) Low attenuation, all-plastic optical fiber. UK Patent GB 2006 790B.
23. DuPont de Nemours & Co (1978) Low attenuation optical fiber of deuterated. UK Patent GB 2007 870B.
24. Mitsubishi Rayon Co (1974) UK Patent 1 431 157.
25. Mitsubishi Rayon Co (1974) UK Patent 1 449 950.
26. Marcuse, D. (1981) *Principles of Optical Fiber Measurements*, Academic Press, New York, pp. 226–232.
27. Kaino, T., Fujiki, M., and Nara, S. (1981) *J. Appl. Phys.*, **52**, 7061–7063.
28. Kaino, T., Jinguji, K., and Nara, S. (1983) *Appl. Phys. Lett.*, **42**, 567–569.
29. Urbach, F. (1953) *Phys. Rev.*, **92**, 1324.
30. Kaino, T., Fujiki, M., and Jinguji, K. (1984) *Rev. ECL*, **32**, 478–488.
31. Yamashita, T. and Kamada, K. (1993) *Jpn. J. Appl. Phys.*, **32**, 2681–2686.
32. Groh, W. (1988) *Makromol. Chem.*, **189**, 2861–2874.
33. Groh, W. and Zimmerman, A. (1991) *Macromolecules*, **24**, 6660–6663.
34. Dettenmaier, M. and Fischer, E.W. (1976) *Makromol. Chem.*, **177**, 1185–1197.
35. Einstein, A. (1910) *Ann. Phys.*, **33**, 1275–1298.
36. Koike, Y., Tanio, N., and Ohtsuka, Y. (1989) *Macromolecules*, **22**, 1367–1373.
37. Tanio, N., Koike, Y., and Ohtsuka, Y. (1989) *Polym. J.*, **21**, 259–266.
38. Kaino, T., Fujiki, M., Oikawa, S., and Nara, S. (1981) *Appl. Opt.*, **20**, 2886–2888.
39. Yamashita, T. and Kamada, K. (1993) *Jpn. J. Appl. Phys.*, **32**, 1730–1735.

40. Yamashita, T. and Kamada, K. (1993) *Jpn. J. Appl. Phys.*, **32**, 4622–4627.
41. Matsuoka, S. and Bair, H.E. (1977) *J. Appl. Phys.*, **48**, 4058–4062.
42. Kyuzyk, M.G. (2007) *Polymer Fiber Optics: Materials, Physics, and Applications*, Taylor & Francis Group, Boca Raton, FL, London, New York, pp. 170–176.
43. Koike, Y., Kimoto, Y., and Ohtsuka, Y. (1982) *Appl. Opt.*, **21**, 1057–1062.
44. Koike, Y., Hatanaka, H., and Ohtsuka, Y. (1984) *Appl. Opt.*, **23**, 1779–1783.
45. Koike, Y. (1991) *Polymer*, **32**, 1737–1745.
46. Koike, Y. (1991) Optical resin materials with distributed refractive index process for producing the materials, and optical conductors using the materials. US Patent 5 541 247, JP Patent 3 332 922, EU Patent 0 566 744, KR Patent 170 358, CA Patent 2 098 604 (originally filed in 1991).
47. Ho, B.C., Chen, J.H., Chen, W.C., Chang, Y.H., Yang, S.Y., Chen, J.J., and Tseng, T.W. (1995) *Polym. J.*, **27**, 310–313.
48. Duijnhoven, F. and Bastiaansen, C. (1999) *Adv. Mater.*, **11**, 567–570.
49. Park, C.-W., Lee, B.S., Walker, J.K., and Choi, W.Y. (2000) *Ind. Eng. Chem. Res.*, **39**, 79–83.
50. Sohn, I.-S. and Park, C.-W. (2001) *Ind. Eng. Chem. Res.*, **40**, 3740–3748.
51. Im, S.H., Suh, D.J., Park, O.O., Cho, H., Choi, J.S., Park, J.K., and Hwang, J.T. (2002) *Appl. Opt.*, **41**, 1858–1863.
52. Villegas, A.G., Ocampo, M.A., Luna-Barcenas, G., and Saldivar-Guerra, E. (2009) *Macromol. Symp.*, **283–284**, 336–341.
53. Koike, Y., Ishigure, T., and Nihei, E. (1995) *J. Lightwave Technol.*, **13**, 1475–1489.
54. Carraher, C.E. (2003) *Polymer Chemistry*, 6th Revised and Expanded edn, Marcel Dekker, Inc., New York, Basel, pp. 309–313.
55. Koike, Y. and Ishigure, T. (2006) *J. Lightwave Technol.*, **24**, 4541–4553.
56. Koike, Y. and Naritomi, M. (1998) Graded-refractive-index optical plastic material and method for its production. US Patent 5 783 636.
57. Asai, M., Hirose, R., Kondo, A., and Koike, Y. (2007) *J. Lightwave Technol.*, **25**, 3062–3067.
58. Hirose, R., Asai, M., Kondo, A., and Koike, Y. (2008) *Appl. Opt.*, **47**, 4177–4185.
59. Daum, W., Krauser, J., Zamzow, P.E., and Ziemann, O. (2002) *Polymer Optical Fibers for Data Communication*, Springer-Verlag, Berlin, Heidelberg, New York, pp. 375–380.
60. Iga, K. (2008) *Jpn. J. Appl. Phys.*, **47**, 1–10.
61. Ziemann, O., Krauser, J., Zamzow, P.E., and Daum, W. (2008) *POF Handbook-Optical Short Range Transmission Systems*, 2nd edn, Springer-Verlag, Berlin, Heidelberg, pp. 330–334.
62. Boutevin, B., Rousseau, A., and Boscz, D. (1992) *J. Polym. Sci., Part A: Polym. Chem.*, **30**, 1279–1286.
63. Koike, K. and Koike, Y. (2009) *J. Lightwave Technol.*, **27**, 41–46.
64. Koike, K., Kado, T., Satoh, Z., Okamoto, Y., and Koike, Y. (2010) *Polymer*, **51**, 1377–1385.
65. Gordon, M. and Taylor, J.S. (1952) *J. Appl. Chem. USSR*, **2**, 493–500.
66. Yamada, B., Sugiyama, S., Mori, S., and Otsu, T. (1981) *J. Macromol. Sci. Chem.*, **A15**, 339–345.
67. Yamada, B., Tanaka, T., and Otsu, T. (1989) *Eur. Polym. J.*, **25**, 117–120.
68. Teng, H., Yang, L., Mikes, F., Koike, Y., and Okamoto, Y. (2007) *Polym. Adv. Technol.*, **18**, 453–457.
69. Giannetti, E. (2001) *Polym. Int.*, **50**, 10–26.
70. Nakamura, M., Kaneko, I., Oharu, K., Kojima, G., Matsuo, M., Samejima, S., and Kamba, M. (AGC) (1990) Novel fluorine-containing cyclic polymer. US Patent 4 897 457.
71. Yamamoto, K. and Ogawa, G. (2005) *J. Fluorine Chem.*, **126**, 1403–1408.
72. Altkorn, R., Koev, I., Van Duyne, R.P., and Litorja, M. (1997) *Appl. Opt.*, **36**, 8992–8998.
73. Tanikkul, S., Jakmunee, J., Rayanakorn, M., Grudpan, K., Marquardt, B.J., Gross, G.M., Prazen, B.J., Burgess, L.W., Christian, G.D., and Synovec, R.E. (2003) *Talanta*, **59**, 809–816.

74. AGC *http://www.lucina.jp/eg_lucina/indexeng.htm* (accessed 1 September 2010).
75. Tanio, N. and Koike, Y. (2000) *Polym. J.*, **32**, 43–50.
76. Kaino, T. and Katayama, Y. (1989) *Polym. Eng. Sci.*, **29**, 1209–1214.
77. Kondo, A., Ishigure, T., and Koike, Y. (2005) *J. Lightwave Technol.*, **23**, 2443–2448.
78. Olshansky, R. and Keck, D.B. (1976) *Appl. Opt.*, **15**, 483–491.
79. Cohen, L.G. (1976) *Appl. Opt.*, **15**, 1808–1814.
80. Fleming, J.W. (1976) *J. Am. Ceram. Soc.*, **59**, 503–507.
81. Cohen, L.G. and Lin, C. (1977) *Appl. Opt.*, **16**, 3136–3139.
82. Giaretta, G. Wegmueller, M., and Yelamarty, R.V. (1999) 11 Gb/sec data transmission through 100 m of perfluorinated graded-index polymer optical fiber. Proceedings of Optical Fiber Communication Conference, 1999, p. PD-14.
83. Polley, A., Balemarthy, K., and Ralph, S.E. (2007) The Optical Society of America/The Conference on Lasers and Electro-Optics, p. CWM5.
84. Polley, A. and Ralph, S.E. (2008) Optical fiber communication conference and exposition/National fiber optic engineers conference. Proceedings OFC/NFOEC, 2008, p. OWB2.
85. Nuccio, S.R., Christen, L., Wu, X., Khaleghi, S., Yilmaz, O., Wilner, A. E., and Koike, Y. (2008) European conference on optical communications, p. We.2.A.4.
86. Intel Co *http://techresearch.intel.com/articles/None/1813.htm* (accessed 1 September 2010).
87. Finisar Co *http://www.finisar.com/cables* (accessed 1 September 2010).
88. AGC *http://www.lucina.jp/eg_fontex/* (accessed 1 September 2010).
89. Noda, T. and Koike, Y. (2010) *Opt. Express*, **18**, 3128–3136.

Index

Fundamentals of Plastic Optical Fibers, First Edition. Yasuhiro Koike.